Strengthening European Energy Policy

"This book provides an insightful perspective on the European energy policy landscape from an unexplored angle: interdisciplinary dialogue. It highlights the crucial role of research and knowledge in policymaking processes and underscores the, often underestimated, potential of creating connections between Social Sciences and Humanities (SSH) and Science, Technology, Engineering, and Mathematics (STEM). This interdisciplinary approach is essential for effectively advancing a low-carbon energy transition."
—Marta Arosio, *Community Manager, Energy Cities, Belgium*

"The book covers a variety of important and stimulating aspects on the European energy transition. By fundamentally appreciating the need to improve the relationship between citizen needs and energy policymaking, the authors develop novel solutions and approaches to tackle the non-technical challenges of building a sustainable future."
—Martin Baumann, *Principal Expert Energy Economics, Austrian Energy Agency, Austria*

"This book delves into the 'how' of the energy transition, examining what's necessary to make it happen and offering concrete solutions to garner support and ownership in a fair and equitable manner. It comes at a crucial time, as the EU just agreed on revamped energy laws, making it essential reading for policymakers and all involved in implementing the transition."
—Christophe Jost, *Senior EU Policy Officer, CEE Bankwatch Network, Belgium*

"While we know how societies could technically go about mitigating climate change, our understanding of how policies could and should make that low-carbon energy transition fair and beneficial is limited. This book offers distinct perspectives on how to reduce this cognitive gap by exploiting synergies between the Social Sciences & Humanities and the Technical & Natural Sciences. It is a concrete basis for further dialogue both between researchers from such diverse fields, as well as with policymakers, on how to smoothly navigate the implementation of EU energy policy goals."
—Maria Kola-Bezka, *Assistant Professor in the Department of Economic Policy & Regional Studies, Nicolaus Copernicus University in Torun, Poland*

"The authors brilliantly navigate the complex landscape of energy transition in Europe. Through insightful topics focusing on community participation, knowledge dissemination, technology delivery, and policy development, this book offers pragmatic solutions for building a sustainable energy future. A must-read for policymakers, researchers, and anyone invested in the European Green Deal's success."
—Wen Liu, *Assistant Professor in the Copernicus Institute of Sustainable Development, Utrecht University, The Netherlands*

"This book is an exciting experiment in bringing together interdisciplinary teams of Social Scientists, Engineers and practitioners to develop recommendations for the governance of the energy transition. What I found particularly appealing with the concise and policy-oriented chapters is that they nicely demonstrate the added value of combining deep technical knowledge on the details of cyberthreats or energy models with Social Science insights into questions such as regulatory contexts or social justice. Instead of broad and general policy advice, we get specific and concrete recommendations grounded in a profound understanding of the problems they aim to address."
—Harald Rohracher, *Professor of Technology & Social Change, Linköping University, Sweden*

"Approaches for addressing climate change, improving energy security, and unlocking the benefits of renewable energy for households and businesses need to move beyond technologies and infrastructures, to consider the politics of delivering a just transition. These approaches need to be shaped by European publics, rather than imposed on them. This book makes an important contribution to the debate about how a just transition can be achieved through the presentation of evidence-based interdisciplinary policy recommendations. By bringing together Technical and Social Science researchers in new collaborations, it goes beyond simple debates between advocates of technological and social solutions."
—Jim Watson, *Professor of Energy Policy, University College London, UK*

Ami Crowther · Chris Foulds · Rosie Robison ·
Ganna Gladkykh
Editors

Strengthening European Energy Policy

Governance Recommendations From Innovative
Interdisciplinary Collaborations

Editors
Ami Crowther
Global Sustainability Institute
Anglia Ruskin University
Cambridge, UK

Chris Foulds
Global Sustainability Institute
Anglia Ruskin University
Cambridge, UK

Rosie Robison
Global Sustainability Institute
Anglia Ruskin University
Cambridge, UK

Ganna Gladkykh
The European Energy Research Alliance
Brussels, Belgium

ISBN 978-3-031-66480-9   ISBN 978-3-031-66481-6 (eBook)
https://doi.org/10.1007/978-3-031-66481-6

© The Editor(s) (if applicable) and The Author(s) 2024. This book is an open access publication.

Open Access This book is licensed under the terms of the Creative Commons Attribution 4.0 International License (http://creativecommons.org/licenses/by/4.0/), which permits use, sharing, adaptation, distribution and reproduction in any medium or format, as long as you give appropriate credit to the original author(s) and the source, provide a link to the Creative Commons license and indicate if changes were made.
The images or other third party material in this book are included in the book's Creative Commons license, unless indicated otherwise in a credit line to the material. If material is not included in the book's Creative Commons license and your intended use is not permitted by statutory regulation or exceeds the permitted use, you will need to obtain permission directly from the copyright holder.
The use of general descriptive names, registered names, trademarks, service marks, etc. in this publication does not imply, even in the absence of a specific statement, that such names are exempt from the relevant protective laws and regulations and therefore free for general use.
The publisher, the authors and the editors are safe to assume that the advice and information in this book are believed to be true and accurate at the date of publication. Neither the publisher nor the authors or the editors give a warranty, expressed or implied, with respect to the material contained herein or for any errors or omissions that may have been made. The publisher remains neutral with regard to jurisdictional claims in published maps and institutional affiliations.

Cover illustration: © Harvey Loake

This Palgrave Macmillan imprint is published by the registered company Springer Nature Switzerland AG
The registered company address is: Gewerbestrasse 11, 6330 Cham, Switzerland

If disposing of this product, please recycle the paper.

Foreword 1: Low-Carbon Approaches at the Crossroads: Why the European Green Deal Will Benefit from Interdisciplinary Insights

Rosalinde van der Vlies *is the Director of the Clean Planet Directorate in the European Commission's Directorate-General for Research and Innovation, and Deputy Mission Manager of the Climate-Neutral and Smart Cities Mission. Before her appointment as Director, Rosalinde van der Vlies was the Head of Coordination & Interinstitutional Relations Unit, and acting Head of Communication & Citizens Unit. Previously, she held positions in Directorate-General Environment, Directorate-General Justice and Home Affairs, and in the private office of Janez Potočnik, the European Commissioner for the Environment. Before joining the European Commission, she worked as a competition lawyer in an international law firm in Brussels and was a part-time teacher at the Catholic University in Brussels.*

The European Union (EU) has outlined its ambitions to become the first climate neutral continent. The achievement of this ambition is supported through the EU Green Deal which sets out a long-term roadmap to deliver on the long-term systemic changes required. The roadmap covers a range of activities across sectors, including climate, energy, and mobility. At the heart of the EU Green Deal is the commitment to put people first and leave no person (or region) behind.

The contribution of Social Sciences and Humanities (SSH) research to low-carbon transitions cannot be understated. I am firmly convinced that we will fail in delivering upon our climate neutrality ambitions without SSH. SSH research contributes to low-carbon transitions in multiple ways, including the development of inclusive approaches, the establishment of effective communication, and the creation of appropriate governance structures.

SSH research supports the establishment of an inclusive approach to achieving the EU's climate neutrality ambitions. As mentioned previously, a key component of the EU's approach to achieving climate neutrality is that no-one, and no region, is left behind. Inclusiveness is therefore important to ensure potential disparities and inequalities are addressed. SSH insights support the establishment of inclusive practices by providing insight into cultural factors, such as values, beliefs, and identities, and how these can support green policies and green transitions. Understanding the different cultures and experiences of individuals is a key component of ensuring no-one is left behind.

SSH research also supports communication and informs the development of effective public engagement initiatives. The incorporation of SSH insights into the narratives of transition can help convey to the public how low-carbon transitions are beneficial for the planet, the health and wellbeing of individuals, and the economy. SSH also help to demonstrate the necessity of behaviour changes to deliver on climate change. It is vitally important that policymakers get support on how to convey messages of urgency—but also the benefits on the lives of individual people—related to low-carbon policies. Not only does SSH research support effective communication related to low-carbon behaviours and practices, but it also provides insights on how changes will both be experienced and encouraged.

SSH research does not focus solely on behaviours and societal configurations; it also provides insights related to policy structures, institutions, and industry. These understandings can inform practices and help ensure that the actions undertaken are as effective as possible.

While SSH research and insights play a vital role in achieving the EU's low-carbon ambitions, they will have the most impact when integrated with other disciplinary perspectives, including those from Science, Technology, Engineering, and Mathematics (STEM). Most solutions for achieving low-carbon ambitions are situated at a crossroads: these solutions are not linked to a single sector or disciplinary background, rather they overlap between sectors and require the integration of different knowledge and perspectives. Bringing together different experts and experiences when developing approaches is essential to find innovative approaches to tackle climate change, undertake the energy transition, and establish sustainable mobility. Yet, in order to achieve all this, there is the need to continue breaking the silos in which research is undertaken and communicated. As such, interdisciplinarity between SSH and STEM needs to be promoted and supported.

Not only is there the need for interdisciplinarity between research disciplines, but there is also the need to have collaboration and communication between different actors. Achievement of low-carbon ambitions requires interactions between SSH researchers, policymakers, and more technical and naturalistic disciplines. Policymaking needs to become more comprehensive and interdisciplinary in order to advance transitions, avoid duplications, and maximise impact through involving different people. The interactions between policy and research are critical as research and innovation activities need to be supported by the regulatory framework, with the regulatory framework also needing to be aware of the research and innovation activities undertaken to enable updates. Within policymaking, we need to continue to break those silos, adopt more interdisciplinary approaches, and make sure to bring along the societal dimension of the transition.

The collective intelligence across the three SSH CENTRE books—bringing together more than 150 researchers from more than 23 countries—is inspiring. The collaborations underpinning the chapters show how we should be working and are a starting point for breaking down silos. I really believe that collaborations between Social Scientists, Humanities researchers, and researchers from more technical disciplines are key to advance low-carbon transitions. In order to achieve climate neutrality, there are many challenges to overcome, but the insights presented within these chapters and the expertise of the chapter authors can support the establishment of effective solutions, help break down barriers, and accelerate pathways to a sustainable and prosperous future.

Brussels, Belgium   Rosalinde van der Vlies

Foreword 2: Beyond Technology: Transcending Disciplinary Boundaries to Achieve More Sustainable Energy Systems

Leen Govaerts *is the Director of Water & Energy Transition at VITO and is a board member of EnergyVille—a research collaboration between the Belgian research partners KU Leuven, VITO, imec and UHasselt in the field of sustainable energy. She is also acting chair of BERA, the Belgian Energy Research Association, and is vice-president of EERA, the European energy research association. At VITO, Leen initiated the programme on international climate services which focuses on establishing international partnerships on climate and sustainable development to support capacity building. She holds a master's degree in business engineering (KULeuven, 1995) and an Executive MBA (Antwerp Management School, 2018).*

I started my journey in energy research as a young Economics graduate at VITO, an applied research and technology organisation, at the end of the twentieth century. It was a time when a wave of Economists came into the research institute to join forces with Engineers, Chemists, and Material Scientists. It swiftly became evident that while we could develop the most advanced technologies imaginable, without the requisite economic rationale and supportive policy framework, widespread market adoption would not happen. This was why I studied Economics and Business Engineering: not only to grasp the basics of engineering but also to comprehend the business dynamics necessary to bring innovations to market and society.

Over the years, the interdisciplinary nature of our research institute has flourished. Sociologists, Political Scientists, Psychologists, Linguists, and Communication Specialists now constitute a significant portion of our institute's expertise, collaborating with Engineers, Chemists, Geologists, ICT Specialists, and other Science, Technology, Engineering and Mathematics (STEM) professionals. Yet the collaboration opportunities are not always obvious. To support collaboration, we need to create a mutual understanding of the sustainability challenges and solutions to navigate together the transitions towards a resource-efficient and climate-resilient economy and society. To support the transition towards a more sustainable economy, there's the need for the different disciplines to understand each other's vocabulary and research instruments.

The imperative role of Social Sciences and Humanities in fostering the transition to sustainable and climate-resilient energy and mobility systems cannot be overstated. Take, for instance, the challenge of enhancing the energy performance of our building stock. While many innovations and technologies are market ready, the pace of renovation lags the pathway needed to achieve net-zero targets by 2050. Community-led energy initiatives offer interesting opportunities for progressing sustainable energy transitions. By investing in local renewable energy production, citizens can drive change, but with the right expertise, ancillary benefits can be harnessed. In my own community, I started an energy cooperative alongside local enthusiasts. Together, we mobilised citizen capital to invest in solar power installations on the roofs of public buildings and local companies. Soon, we realised that we should go further to increase our impact. Collaborating with social workers, the cooperative expanded its activities to encompass renewables and other renovation measures for socially disadvantaged households, empowering them to access renewable energy benefits and contribute to a just transition. The energy transition will not be sustainable if only those who have the capital can benefit from the financial and comfort gains it brings.

Yet, the significance of Social Sciences and Humanities extends far beyond technology acceptance and adoption. We are the first generation to confront the tangible impacts of climate change and the last with the capacity to mitigate global warming. Achieving a low-carbon energy transition is challenging and complex as technologies are, and will continue to be, intertwined with everyday life—for example, locally-produced renewable energy, digitalised smart appliances, smart grids to accommodate flexibility, and new high-voltage lines bringing offshore electricity production inland.

Translating this complexity into clear, accessible evidence-based insights is a priority. In these tumultuous times, marked by geopolitical conflicts, energy crises, and climate disasters, citizens often exhibit risk aversion and resistance to change. Conservative populist voices, thriving on fake news and propagating climate scepticism, gain traction and reduce support for sustainability transitions, for example by attributing unaffordable energy prices to initiatives like the European Green Deal. Engaging citizens in positive thinking about the sustainability transitions and persuading them to invest in the present and future well-being of generations to come is paramount. Social Sciences and Humanities research can have a substantial contribution to support positive thinking and engaging citizens.

I have always championed the transformative power of technology in shaping a sustainable and equitable world. However, to ensure its impact in enhancing the well-being of global citizens, we must draw upon the creative insights of Social Sciences and Humanities. I extend my gratitude to the authors for their contributions to interdisciplinary discourse and hope that this book serves as a catalyst for researchers and sustainability practitioners to transcend the borders of their disciplines and collaboratively forge a brighter future.

Antwerp Metropolitan Area   Leen Govaerts

Preface

How can research support better energy policy in Europe, for a fair, reliable and low-carbon energy future? This book is built on the premise that energy issues are never *only* social or *only* technical in nature, but instead require researchers from Computer Science to Sociology having the tools to be able to work together effectively. While *Social Sciences and Humanities* (SSH) energy research has received less funding to date than *Science, Technical, Engineering, and Mathematics* (STEM) energy research, as we get to the sharp end of trying to implement policies to meet challenging carbon targets, there is growing recognition that social challenges and solutions need to be better embedded in technical recommendations, and vice versa. The importance of improving SSH-STEM interdisciplinary practice therefore cannot be understated.

As editors, we have had a range of interdisciplinary energy journeys spanning social and technical disciplines. These include: moving from Environmental Sciences to critical Environmental Social Sciences, as part of becoming disillusioned by notions of objectivity and magic-bullet technologies; making a radical shift from a PhD in Applied Mathematics to undertaking qualitative research on the emotional undercurrents of climate (in)action; specialising in Human Geography before starting to engage with other Social Sciences perspectives; and transitioning from Economics to Systems Science and interdisciplinary energy systems modelling/analysis. Indeed, our experiences have shown us there can be just as much variety (and epistemological disagreement!) within SSH or within STEM as between one and the other.

This book is a core output from the Horizon Europe project SSH CENTRE: *Social Sciences and Humanities for Climate, Energy, and Transport Research Excellence*. Each chapter represents findings from a novel collaboration between the social and technical sciences which was catalysed and funded by the SSH CENTRE. These experiments in collaboration build on a legacy of previous initiatives which have sought to strengthen links across and within SSH disciplines (described further in section 1.3), with the inclusion of STEM as co-lead for every chapter being a new ingredient for this initiative.

We wanted to make the call for chapters appealing to both SSH and STEM researchers, and to help with this asked individuals from both communities to review the call text. Even this part of the process highlighted how key terms are interpreted, and used, differently across disciplines. This is just one of the challenges of conducting interdisciplinary research, with others including framing the problem, and determining the methodology. The policy recommendations in this book have been produced by collaborative teams that have bridged disciplinary boundaries—this has involved significant efforts and we greatly appreciate the work that has been undertaken.

Fundamentally, this book is aimed at *Strengthening European energy policy* through better interdisciplinary research. It is part of a three-volume collection; the other volumes focus on recommendations for climate policy and mobility policy. All three are available open access.

Cambridge, UK	Ami Crowther
Cambridge, UK	Chris Foulds
Cambridge, UK	Rosie Robison
Brussels, Belgium	Ganna Gladkykh

Acknowledgements

We thank all those who contributed to this book, especially the chapter teams (Chapters 2–11) who collaborated with colleagues that they had not published with before, and ultimately tried out something new across disciplinary boundaries. We also thank the Foreword and Afterword authors whose engagement with the book, its topics, and its overarching ambition has supported the dialogue we are hoping to stimulate.

We are grateful to several colleagues, whose interesting conversations and insights fed into the development of the book: Melanie Rohse, Davide Natalini, Lara Houston (all Anglia Ruskin University); Ruth Mourik, Marten Boekelo, Razia Jaggoe (all Duneworks); Mojca Drevensek (Consensus Communications); Alevgul Sorman, Ester Galende (both BC3); as well as the editorial teams of our accompanying climate and mobility books, plus the whole SSH CENTRE consortium.

The Editors' time on this book—in addition to the collaboration expenses of the chapter teams—was funded by the SSH CENTRE project. This project is funded by the European Union's Horizon Europe research and innovation programme (grant agreement no. 101069529) and by the UK Research and Innovation under the UK Government's Horizon Europe funding guarantee (grant no. 10038991).

We are grateful for the internal Anglia Ruskin University support from Megan Plumb and Emma Milroy. We also thank Iva Tajnšek

(Consensus Communications) for her assistance in managing the accompanying Zenodo resources. Finally, we thank Rachael Ballard of Palgrave Macmillan for her efficiency and guidance throughout the proposal and manuscript preparation processes.

Contents

Foreword 1: Low-Carbon Approaches at the Crossroads: Why the European Green Deal Will Benefit from Interdisciplinary Insights   v
Rosalinde van der Vlies

Foreword 2: Beyond Technology: Transcending Disciplinary Boundaries to Achieve More Sustainable Energy Systems   ix
Leen Govaerts

Part I Introduction

1 Interdisciplinary Collaborations for European Energy Policy and Governance   3
   Ami Crowther, Chris Foulds, Rosie Robison, and Ganna Gladkykh

Part II Navigating Community Participation

2 Simplify the Uptake of Community Energy by Leveraging Intermediaries and the Use of Digital Planning Tools   17
   Franziska Mey, Kristian Borch, Stephan Bosch, Benita Ebersbach, Robert Hecht, Lars Holstenkamp, and Jörg Radtke

3 Prioritise Inclusive, Early, and Continuous Societal Engagement to Maximise the Benefits of Geothermal Technologies 31
Melanie Rohse, Amel Barich, Claire Bossennec, Annick Loschetter, Adele Manzella, Anna Pellizzone, Stacia Ryder, and Iain Soutar

4 Create a Co-learning Environment for Geothermal Energy Communities Across the European and African Unions 45
Chris Büscher, Walter Wheeler, Susan Onyango, Jacques Varet, Fabio Iannone, Eleonora Annunziata, Yves Geraud, and Peter Omenda

Part III Navigating Knowledges for the Built Environment

5 Facilitate the Development of Energy Literacy Amongst Citizens to Support Their Meaningful Participation in the Energy Transition 61
Philippa Calver, Ami Crowther, and Claire Brown

6 Support Place-Based and Inclusive Supply Chain, Employment and Skills Strategies for Housing-Energy Retrofit 73
Rachel Macrorie, Hadi Arbabi, Will Eadson, Richard Hanna, Kaylen Camacho McCluskey, Kate Simpson, and Faye Wade

Part IV Navigating the Delivery of New Technology

7 Promote Integrated Policy Design to Overcome Social and Technical Challenges for Agrivoltaic Deployment 89
Alessandro Sculio, Pınar Derin-Güre, Ivan Gordon, Angela Ciotola, and Hanna Dittmar

8	Increase Social Acceptability of Nuclear Fusion, Agrivoltaics, and Offshore Wind Through National Support Programmes Pascal Clain, Insaf Khelladi, Christophe Rodrigues, Alessandro Biancalani, Guillaume Guerard, and Saeedeh Rezaee Vessal	101
9	Protect the EU's Digital Energy Infrastructure Against Cyberthreats Through Advanced Technologies, Human Vulnerability Mitigation, and Ethical Practices Amal Mersni, Aliaksandr Novikau, Marcin Koczan, and Abdulfetah Abdela Shobole	115

Part V Navigating Models for Policy Development

10	Understand Stakeholder Perceptions and Implementation Possibilities for Energy Efficiency Measures and Policy Through Multicriteria Modelling Alexandra Buylova, Aron Larsson, Naghmeh Nasiritousi, and Afzal S. Siddiqui	131
11	Rethink Energy System Models to Support Interdisciplinary and Inclusive Just Transition Debates Diana Süsser, Connor McGookin, Will McDowall, Francesco Lombardi, Lukas Braunreiter, and Stefan Bouzarovski	145

Part VI Conclusion

12	Reflections on Interdisciplinary Collaborations for European Energy Policy and Governance Ami Crowther, Chris Foulds, Rosie Robison, and Ganna Gladkykh	161

Afterword 1: A Quest for More Intentional Interdisciplinary Synergies 169
Giulia Sonetti and Osman Arrobbio

Afterword 2: Considering the Role of the Scientific Community 173
Henry Jeffrey and Kristofer Grattan

Afterword 3: Empowering the Energy Transition: Collaborative Pathways Ahead for European Union Policy 177
Emma Bergeling

Afterword 4: Reinserting the Missing Piece: Integrating the Human Dimension in Energy Policy 181
Rod Janssen and Audrey Nugent

Index 185

Notes on Contributors

Eleonora Annunziata is an Associate Professor of Management, focusing on sustainability management at Scuola Superiore Sant'Anna, Pisa. Co-leader of 'Energy and Resource Efficiency' at the Laboratory of Sustainability Management.

Hadi Arbabi is a Lecturer in the Built Environment at the University of Sheffield. Hadi's research focuses on urban systems and infrastructure in the context of decarbonisation and engineering resilience.

Amel Barich is a Senior Project Manager at the Geothermal Research Cluster, Iceland. Geoscientist by background, her research currently focuses on the Social Licence to Operate and communication in geothermal energy.

Alessandro Biancalani is an Associate Professor of Physics at ESILV Engineering School and DVRC lab member. His research focuses on investigation of waves and instabilities in magnetised plasmas in magnetic-confinement nuclear fusion devices.

Kristian Borch is an Associate Professor at Aalborg University, Denmark, and Senior Scientist at Ruralis, Norway with a background in Biotechnology, investigating just energy transition, public acceptance, and fair distribution.

Stephan Bosch is an Energy Researcher with a background in Geography at the University of Augsburg, focusing on energy transitions, spatial planning, energy landscapes, energy justice, and geographical information systems.

Claire Bossennec holds a PhD in Geoscience, a Researcher at the Helmholtz-Zentrum Potsdam Deutsches GeoForschungsZentrum GFZ, specialises in thermal storage and geothermal projects implementation at different stages, from feasibility studies, exploration, and construction.

Stefan Bouzarovski is a Professor of Human Geography at the University of Manchester, and a Senior Expert at IEECP. His research interests focus on energy injustices, housing, and urban development.

Lukas Braunreiter works at the Swiss Energy Foundation. Before, he worked at Zurich University of Applied Sciences on plausible, desirable energy futures. His background is in Environmental Science and policy.

Claire Brown is an Environmental Management practitioner and PhD Researcher at the Tyndall Centre for Climate Change Research at the University of Manchester. Her research focuses on housing and energy.

Chris Büscher is a Postdoctoral Research Fellow, Department of Cultures, Politics, and Society, University of Turin. He focuses on water and energy politics in Africa, drawing on Political Ecology and Development Studies.

Alexandra Buylova is a Research Fellow in Political Science at the Swedish Institute of International Affairs. Her current research focuses on climate governance and politics.

Philippa Calver has a PhD in Engineering and is a Lecturer in Sustainability at the Global Sustainability Institute at Anglia Ruskin University. Her research focuses on justice within societal transitions.

Kaylen Camacho McCluskey is a Research Assistant at the Energy Futures Lab, Imperial College London, preparing a policy briefing paper on energy sector skills and training requirements in the UK.

Angela Ciotola is a PhD candidate at the Karlsruhe Institute of Technology. With a background in Chemical Engineering, she specialises in risk engineering and is involved in analysing the risks of the large-scale deployment of solar technologies.

Pascal Clain is an Assistant Professor of Chemical Engineering at ESILV Engineering School and DVRC lab member. His research focuses on developing materials and processes for eco-friendly thermal energy storage in refrigeration.

Ami Crowther is a Postdoctoral Research Fellow at the Global Sustainability Institute, Anglia Ruskin University. She is a Human Geographer interested in the governance of place-based (urban) energy transitions.

Pinar Derin-Güre is an Associate Professor of Economics at Middle East Technical University. She is a member of EERA e3S and is the coordinator of EERA PV Subprogramme 5: Socio-economic aspects of Photovoltaics.

Hanna Dittmar is a Project Officer at SolarPower Europe. She is also part of the Secretariat of the European Technology and Innovation Platform, coordinating the Social PV Working Group that is targeting the socio-economic issues of PV deployment.

Will Eadson is a Professor of Urban and Regional Studies at Sheffield Hallam University, where he leads an international research programme on green and just economic development.

Benita Ebersbach is a Research Associate at RIFS holding a M.Sc. in Psychology with her research focusing on participation in energy transitions and specific interest in individual behavioural change.

Chris Foulds is a Professor of Sustainability & Society, at Anglia Ruskin University's Global Sustainability Institute, and co-lead of SSH CENTRE. He is an Environmental Social Scientist exploring governance for sustainability.

Yves Geraud is a Professor at the National School of Engineering Geology, and head of the Geology-Energies team, at the GeoRessources Laboratory, University of Lorraine, with a focus on geo-energetic systems characterisation.

Ganna Gladkykh is a Policy Fellow at the Stockholm Environment Institute. She has a background in Economics and System Dynamics, and previously worked as a Clean Energy Transition Expert at EERA.

Ivan Gordon is the Department Director of IMOMEC at IMEC, the Vice Director of IMO at Hasselt University, and a Professor of Digital Photovoltaics at TU Delft. He is the coordinator of the EERA JP PV.

Guillaume Guerard is an Assistant Professor of Computer Sciences at ESILV Engineering School and DVRC lab member. His research focuses on modelling complex systems, notably Smart Grids and agrivoltaics.

Richard Hanna is a Research Associate at Imperial College London, conducting international reviews for the UK Energy Research Centre on best practices in building energy renovation and heat decarbonisation policies.

Robert Hecht has a background in Cartography and Geoinformatics, working at the Leibniz Institute of Ecological Urban and Regional Development with expertise in spatial modelling, cartography, and remote sensing.

Lars Holstenkamp is an Economist, co-leading the junior research group "SteuerBoard Energie" at Leuphana University Lüneburg, and head of ECOLOG Institute, with a research focus on governance and financing of the energy transition.

Fabio Iannone is a Research Fellow in Management at the Laboratory of Sustainability Management, Scuola Superiore Sant'Anna, Pisa, interested in renewable energy communities, social acceptance, and social impacts of digital/green transitions.

Insaf Khelladi is an Associate Professor of Marketing at EMLV Business School and DVRC lab member, interested in cognitive processes of human-technology interactions and their effects at individual, organisational, and institutional levels.

Marcin Koczan is an Assistant Professor at the Institute of International and Security Studies at the University of Wrocław. His research focuses on Energy Security and Policy.

Aron Larsson is a Professor at the Risk and Crisis Research Centre at Mid Sweden University, and an Associate Professor at Stockholm University. He researches risk and decision analysis, and simulation for resilience assessment.

Francesco Lombardi is an Assistant Professor at TU Delft. He is an Energy Engineer, working on computational methods for the design of socially-just and technically-robust energy systems across scales.

Annick Loschetter is an Engineer and Project Manager at BRGM in Applied Geosciences research, working on CO_2 geological storage and Geothermal Systems. She is experienced in scientific animation and interdisciplinary facilitation.

Rachel Macrorie is a Postdoctoral Research Associate at Sheffield Hallam University. An Urban and Environmental Geographer, her research examines the governance of sustainability transformations across diverse urban domains, including retrofit.

Adele Manzella is a Senior Scientist at the National Research Council of Italy. A Geophysicist with experience in geothermal exploration, she also focuses on the social aspects of geothermal energy.

Will McDowall is an Associate Professor in Innovation and Sustainability at the UCL Institute for Sustainable Resources, University College London. He is a Social Scientist with an interdisciplinary background.

Connor McGookin is a Postdoctoral Research Fellow at the Simon Fraser University. He is an Energy Engineer interested in the bridging of energy systems modelling and Social Sciences research.

Amal Mersni is an Assistant Professor in the Department of Engineering at the International University of Sarajevo. Her research focuses on routing, QoS, cyber resilience, network security, and fault tolerance.

Franziska Mey leads the research group Energy Transitions and Public Policy at RIFS. She is a Political Scientist investigating policies and politics, local governance, and justice in energy transitions.

Naghmeh Nasiritousi is an Associate Professor in Political Science, a Senior Lecturer and co-director of the Centre for Climate Science and Policy Research at Linköping University, and a Research Fellow at the Swedish Institute of International Affairs.

Aliaksandr Novikau is an Associate Professor in the Department of Political Science and International Relations at the International University of Sarajevo. His research focuses on Energy Security and Environmental Policy.

Peter Omenda is a Researcher and Consultant with Scientific and Engineering Power Consultants, Nairobi. A background in Geoscience and Engineering, with interests including geothermal energy resources, advanced geothermal systems, and low-temperature utilisation.

Susan Onyango is a Political Anthropologist and Consultant, and a Doctoral Student at Laboratoire d'Anthropologie Politique, L'École des hautes études en sciences sociales, Paris. Interests include gender perspectives in geothermal development along Africa's Rift.

Anna Pellizzone is a freelance Consultant in Responsible Research and Innovation (RRI). A Natural Scientist by training, she designs and runs participatory processes in RRI as a stakeholder engagement professional.

Jörg Radtke is a Political Scientist and Senior Researcher at RIFS. His research concentrates on governance, politics, and policies, as well as participation and conflicts in the German energy transition.

Saeedeh Rezaee Vessal is an Associate Professor of Marketing at EMLV Business School and DVRC member. Her research applies Behavioural Sciences methods to explore responsible consumption, human-technology interactions, and individuals' well-being.

Rosie Robison is a Professor of Social Sustainability at Anglia Ruskin University, and co-leads the SSH CENTRE. Working across Psychosocial Studies and Geography, she also holds a PhD in Applied Mathematics.

Christophe Rodrigues is an Assistant Professor of Computer Sciences at ESILV Engineering School and DVRC lab member. His areas of expertise are machine learning and applications of natural language processing.

Melanie Rohse is an Associate Professor at the Global Sustainability Institute, Anglia Ruskin University. An Environmental Social Scientist, she researches everyday experiences with energy, including engagement in energy systems.

Stacia Ryder is an Assistant Professor of Sociology at Utah State University. Her research focuses on power, place and participation in environmental, energy, and climate justice contexts.

Alessandro Sculio is an Assistant Professor at the University of Torino. He is the coordinator of the European Energy Alliance (EERA) Clean Energy Transition for Sustainable Society (e3s) joint programme (JP).

Abdulfetah Abdela Shobole is an Assistant Professor in the Department of Electrical Engineering at Istanbul Sabahattin Zaim University.

His research focuses on smart grid protection, cybersecurity, and smart grid automation.

Afzal S. Siddiqui is a Professor of Computer and Systems Sciences at Stockholm University and an Adjunct Professor in the Department of Mathematics and Systems Analysis at Aalto University.

Kate Simpson is an Associate Professor, Centre for Sustainable Construction and Retrofit, Nottingham Trent University. Kate's socio-technical research on housing retrofit includes practices, processes, experiences, and outcomes, including capabilities and motivations.

Iain Soutar is a Senior Lecturer in Energy Policy at the University of Exeter. He is an Environmental Social Scientist focusing on processes of engagement and participation in energy innovation projects.

Diana Süsser is a Senior Expert at the Institute for European Energy and Climate Policy (IEECP). She is an interdisciplinary Geographer interested in participatory modelling for just transition pathways.

Jacques Varet is a Geothermal Expert and CEO of Géo2D. A Volcanologist, currently focusing on the East African Rift System. Prior leader of BRGM (French Geological Survey), Eurogeosurvey, French Environment Institute.

Faye Wade is a Chancellor's Fellow at the University of Edinburgh. Faye uses a Sociological approach to understand the working cultures of supply chain professionals responsible for delivering energy retrofit.

Walter Wheeler is a Geoscientist focusing on extensional and strike-slip systems, geothermal, fluid flow pathways, and community-scale integrated energy systems. A Senior Researcher in the Energy & Technology Department, NORCE Norwegian Research Centre.

LIST OF TABLES

Table 2.1	Definitions of community-led energy initiatives by the European Commission (Inspired by European Union, 2023)	20
Table 4.1	Comparison of renewable energy sources, highlighting the strengths of each, and aspects unique to geothermal	49
Table 6.1	Examples of supply chain support in One-Stop Shops	78
Table 9.1	Analysis of key EU cybersecurity legislation specifically for the digital energy sector	123
Table 11.1	How Social Sciences can contribute to filling modelling gaps on energy justice	149
Table 12.1	Chapter policy recommendations and related EU policy instrument(s)	163

PART I

Introduction

CHAPTER 1

Interdisciplinary Collaborations for European Energy Policy and Governance

Ami Crowther, Chris Foulds, Rosie Robison, and Ganna Gladkykh

Abstract The European Union (EU) has set out ambitious targets to address European sustainability and environmental challenges. As part of this, policies and initiatives focusing on both energy supply and energy demand have been established. Yet, the complexity of enacting the required actions demonstrates the need for interdisciplinary collaboration to inform approaches. The integration of Social Sciences and

---

A. Crowther (✉) · C. Foulds · R. Robison
Global Sustainability Institute, Anglia Ruskin University, Cambridge, UK
e-mail: ami.crowther@aru.ac.uk

C. Foulds
e-mail: chris.foulds@aru.ac.uk

R. Robison
e-mail: rosie.robison@aru.ac.uk

G. Gladkykh
European Energy Research Alliance, Brussels, Belgium
e-mail: ganna.gladkykh@sei.org

© The Author(s) 2024
A. Crowther et al. (eds.), *Strengthening European Energy Policy*,
https://doi.org/10.1007/978-3-031-66481-6_1

Humanities (SSH) and the technical sciences (Science, Technology, Engineering and Mathematics (STEM)) provides opportunities to identify policy approaches which challenge the status quo, supporting low-carbon energy transitions. This book includes interdisciplinary recommendations for relevant EU energy strategies—a total of 10 recommendations, split into four parts, are presented. This chapter includes a breakdown of the book's chapters and its overall narrative. The chapter closes with tips on how to read the book as a whole, as well as the individual chapters it is composed of.

Keywords Interdisciplinarity · Dialogue · Collaboration · EU policy · Research & innovation

1.1 Current EU Energy Policy Ambitions

Over the last 10–20 years, the European Union (EU) has become increasingly interested in decarbonising its energy system. Whilst we are not going to comprehensively narrate the evolution of EU energy policy across this timeframe—such as unpacking the relationship between current policies and their predecessors, e.g. EU Energy Union (European Commission, 2015) and Clean energy for all Europeans (European Commission, Directorate-General for Energy, 2019)—we do feel it worthwhile to articulate the diversity of policies established by the EU to support low-carbon energy transitions. The EU is a global leader in energy policymaking, with this providing multiple opportunities and avenues through which researchers can engage with energy policy.

The flagship EU Green Deal is a set of policy programmes that address an array of European sustainability and environmental challenges—within this, updated commitments have been made to drive progress on both energy demand and supply matters. For example, the EU Green Deal triggered the development of the: European Climate Law (European Union, 2021), which established a framework for the EU achieving carbon neutrality by 2050; and, the Renovation Wave Strategy (European Commission, 2020), which, set out to reduce barriers to energy and resource-efficient renovation, and included, for example, ambitions to at least double the annual energy renovation rate of residential and non-residential buildings by 2030. EU Green Deal commitments also led to the 'Fit for 55' package of legislation, which primarily involved a range of updates to existing energy-related EU Directives, such as Energy Efficiency Directive (European Union 2023a); European Performance of Buildings Directive (European Union, 2024); and Renewable Energy Directive (European Union, 2023b).

This ambitious work has been built on further, following Russia's invasion of Ukraine, which heightened European concern around energy security. Indeed, this led to the introduction of REPowerEU (European Commission, 2022), which focused on increasing the resilience of the energy system within Europe, and included actions to diversify energy supply and accelerate the clean energy transition.

As shown here, policies are continually built upon with new nuances, in response to developments within the EU and shifts in the geopolitical landscape. The pace of policy development and the urgency in which those policies need to be delivered on-the-ground is also clear. We take the view in this book that the achievement of stated ambitions requires more than the adoption of new technologies; rather, there is a need to rethink how the energy system operates. As part of this, governance approaches should accommodate multiple stakeholders, address conflicting interests, and foster integration between countries. It is exactly in these ways that calls for interdisciplinarity come to the fore.

1.2 The Need for Interdisciplinary Collaboration to Meet Complex European Energy Challenges

When this book refers to 'interdisciplinarity', we are referring to the integration of Social Sciences and Humanities (SSH) and Science, Technology, Engineering, and Mathematics (STEM) perspectives. This is not to say that interdisciplinarity does not occur within SSH or within STEM fields (and is not of value), but the scope of this book firmly focuses on what can be achieved through SSH and STEM working together. SSH disciplines[1] include Sociology, Psychology, Human Geography, Law, Philosophy, etc.; and STEM disciplines include Physics, Computer Science, Civil Engineering, Climate Science, Geology, etc.

---

[1] See Foulds and Robison (2018, pp. 3–4) for a distinction between the energy-related Social Sciences and energy-related Humanities.

Increasingly, European policymakers are calling for the research and innovation (R&I) they fund to better integrate STEM and SSH. Indeed, European Commission commitments to 'mainstreaming SSH' (i.e. embedding SSH within the proposal template) or 'flagging SSH' (i.e. labelling calls that are particularly relevant for SSH perspectives) have become embedded in the EU's recent Framework Programmes. However, we argue that SSH is often not taken seriously by STEM researchers running EU-funded projects. More broadly, policymakers often have a preference for STEM-led knowledge, which is deemed to be more objective, quantitative, instrumental, and technologically-grounded (Royston & Foulds, 2021).

Therefore, despite attempts to have more interdisciplinary R&I, STEM perspectives continue to dominate the evidence base behind Europe's low-carbon energy transition. Across EU Horizon2020, SSH partners received 16% of funding in energy R&I projects (European Commission, Directorate-General for Research and Innovation, 2023). Indeed, the European Commission's approach to innovation is fundamentally technological; for example, social innovation is backgrounded by the prioritisation of Technological Readiness Levels. All this means that much SSH-STEM collaboration can be weak, tokenistic, and/or overtly mainstream, to thus align with dominant R&I expectations. SSH has traditionally played a subordinate role to STEM (Kropp, 2021; Silvast & Foulds, 2022).

Without the meaningful integration of a range of SSH perspectives, energy-related R&I will continue to lack due attention to, for example, social practices, social values, institutional dynamics, materiality, and social justice (Foulds & Robison, 2018; Ingeborgrud et al., 2020). SSH perspectives can support the shift to alternative energy governance, (infra)structures, and transitions, by providing insights into social phenomena that organise how people interact with the energy system, and by incorporating questions of equity and fairness (Foulds & Robison, 2018). These insights are increasingly pertinent as low-carbon energy transitions provide opportunities to radically rethink energy systems and practices, for example by providing opportunities to increase the number of actors participating in the energy system, support energy democratisation, and expand distributed energy generation.

1.3 Stimulating Novel Interdisciplinary Collaborations Through This Book Project

Meeting the EU's energy policy ambitions will require new policy approaches, which may significantly challenge the status quo. Thus, truly innovative interdisciplinary research, that integrates SSH and STEM disciplines, could play a fundamental role in Europe achieving its energy policy ambitions. Through this book project, we wanted to create a funded opportunity for such novel interdisciplinary collaborations. As such, the aim of this book project is to generate concrete interdisciplinary recommendations for relevant EU energy strategies, by bringing social and technical disciplines together on a more equal footing.

In doing so, this book looks *beyond* the issue of SSH fragmentation, to focus on interdisciplinary SSH-STEM collaborations. SSH fragmentation, i.e. the challenge of bringing together researchers from different SSH disciplines, has been addressed in-depth in previous initiatives we have led. These initiatives include the open access book, *Advancing Energy Policy* (Foulds & Robison, 2018), which brought together 50 researchers across a variety of SSH disciplines for the SHAPE ENERGY project. Further, we ran four Energy-SHIFTS Working Groups that sought to include a wider diversity of SSH expert voices in setting future R&I agendas for Renewables (Krupnik et al., 2022), Smart Consumption (Robison et al., 2023), Energy Efficiency (Foulds et al., 2022), and Mobility (Ryghaug et al., 2023).

By expanding upon this previous work on SSH fragmentation, to focus on developing SSH-STEM collaborative practices for EU-level energy policy, this book has a dual-purpose: firstly to showcase and disseminate important policy-relevant recommendations; but secondly to understand processes of cross-fertilisation between the SSH-STEM collaborators, and how these processes can support the development of policy recommendations.

1.4 Overview of the Four Parts of the Book and Constituent Chapters

The chapters of this edited collection have been grouped into four Parts: (1) Navigating community participation; (2) Navigating knowledges for the built environment; (3) Navigating the delivery of new technologies; and, (4) Navigating models for policy development. These Parts were not pre-defined, rather they emerged from the contributions from the different chapter teams. The emergent nature of these themes, the intention to start dialogue, and the complexity of interdisciplinary work are why the Parts focus on 'navigation'—in that, the chapters within each of the Parts are not concrete, rather there is a flexibility and an evolving nature to policy recommendations depending upon the broader context they are situated in.

Part I on *Navigating community participation* includes chapters that focus on how community engagement with renewable energy projects and technologies can support positive social outcomes. The chapters posit that citizens need to be engaged in energy projects in order to achieve the EU's low-carbon ambitions, and set out approaches that give communities the opportunity to share their views.

Mey et al. (Chapter 2) call for greater support to be provided for community-led renewable energy initiatives, including both governance mechanisms and digital planning tools. They argue that the leveraging of different instruments at different scales will help unlock the potential of community energy in the EU. Both Rohse et al. (Chapter 3) and Büscher et al. (Chapter 4) focus on the opportunities for geothermal energy in achieving the EU's low-carbon ambitions. Rohse et al. (Chapter 3) recommend greater societal engagement in relation to geothermal developments, including at the project level, but also by regional and EU policymakers and geothermal operators. Whereas Büscher et al. (Chapter 4) focus more on the actions that can be undertaken to support the establishment of geothermal energy communities, referencing the need to involve local people in decision-making processes and facilitating proactive behaviour to support the implementation of energy communities both in the EU and in Africa.

Part II on *Navigating knowledges for the built environment* includes chapters focusing on the knowledge required to achieve the EU's energy ambitions, highlighting different knowledges required to participate in different aspects of the energy transition. The chapters reflect on the importance of how knowledge is framed and shared. Knowledge is considered critical for supporting meaningful and informed participation in the energy system.

Calver et al. (Chapter 5) focus on energy literacy and the need to ensure that citizens have the knowledge to meaningfully participate in the energy system. The chapter sets out different elements and aspects that would support citizen comprehension. Whilst Macrorie et al. (Chapter 6) comment on the need for (re-)training programmes to ensure a diverse, skilled workforce able to support the achievement of EU and National retrofit targets. As part of this, they set out the opportunities of situating knowledge within defined contexts and the establishment of appropriate regulation and training.

Part III on *Navigating the delivery of new technologies* includes chapters on how policies are designed to support shifting energy systems. These chapters outline constituent policies that support both the delivery of new renewable energy technologies, and the digital infrastructures that are in themselves a new form of technology (whilst assisting the rollout of other technologies).

Derin-Güre et al. (Chapter 7) focus on agrivoltaic technologies and the need for an integrated policy framework to support the deployment of this technology. They argue that definitions, funding, and public engagement are critical elements. Clain et al. (Chapter 8) comment on the importance of public social acceptability for the rollout of renewable energy technologies. As part of this, they set out actions that can be undertaken to facilitate public acceptance. Mersni et al. (Chapter 9) call for greater consideration of how digital energy infrastructures can be safeguarded against cyberthreats. Approaches to support this include training programmes and the use of Artificial Intelligent (AI) technologies.

Part IV on *Navigating models for policy development* includes chapters which argue that models can support decision-making and inform policymaking practices, to help achieve EU energy policy ambitions. The chapters outline how models can integrate SSH insights and different perspectives, as well as the opportunities of doing so.

Buylova et al. (Chapter 10) demonstrate the opportunities for multi-criteria models to support decision-making in relation to energy infrastructures. By providing the space for stakeholders to comment on the model, they give insight into how and when the model could be used to support decision-making. Süsser et al. (Chapter 10) reflect on the need to better incorporate SSH perspectives and diverse voices into the development of energy models, adopting a justice lens to frame the discussions. In doing so, they hope to inform the actions of policymakers and incorporate alternative understandings alongside the models.

1.5 Tips on How to Read This Book

The book has been structured to be accessible to a diverse audience. We hope that the book and its contents will support dialogue on energy policy, within and across SSH and STEM disciplines, and also between academics and policymakers.

The book's chapters are intentionally short, with each chapter having a clearly defined purpose and message. Each of the 10 'core' chapters of the book (Chapters 2–11) presents a policy recommendation for EU energy policy and sets out the evidence used to develop the recommendation. As the purpose of these 10 chapters is to introduce the policy recommendation, the chapters do not include a detailed literature review or methodology. Similarly, references are used sparingly to support or reinforce the arguments being made. Some of the chapters reference additional materials that support their chapter—these additional materials have been uploaded to the SSH CENTRE's Zenodo site. The conclusion chapter (Chapter 12) provides reflections on how the recommendation chapters engage with EU energy policy and the interdisciplinary collaborations that informed the recommendations.

There are also structural consistencies across the policy recommendation chapters (Chapters 2–11) to support the accessibility of the book and to help stimulate dialogue across, and between, disciplines and actors. Each of the chapter titles are the policy recommendations that the chapters discuss; we hope that this will immediately communicate the headline messages to the readers from the point of the Contents page onwards. Moreover, at the start of each of the policy recommendation chapters are a series of 'policy highlight' bullets. These bullet points summarise how the chapter's overall policy recommendation can be achieved and reference the interdisciplinary activities that informed the recommendation's development. The conclusion section of the policy recommendation chapters expands upon each of these bullet points. Thus, we hope it is possible to get an overarching understanding of the policy recommendation by reading only the policy highlights at the start and the conclusions at the end.

The Forewords and Afterwords that book-end this book also facilitate dialogue on STEM-SSH (and wider) collaborations for low-carbon energy futures in Europe. The Forewords present the perspectives of two invited individuals, reflecting upon the importance of SSH research and SSH-STEM collaboration. The Afterwords have been written by SSH researchers (Afterword 1), STEM researchers (Afterword 2), a policy actor (Afterword 3), and members of the SSH CENTRE project's Business Advisory Board (Afterword 4) who reflect upon the policy recommendations presented within the book, situating their reflections within their experiences and understandings.

REFERENCES

European Commission. (2015). *A framework strategy for a resilient energy union with a forward-looking climate change policy*. COM/2015/080 final. Brussels: European Commission.

European Commission. (2020). *A renovation wave for Europe—Greening our buildings, creating jobs, improving lives*. COM(2020) 662 final. Brussels: European Commission.

European Commission. (2022). *REPowerEU Plan*. COM(2022) 230 final. Brussels: European Commission.

European Commission, Directorate-General for Energy. (2019). *Clean energy for all Europeans*. Luxembourg: Publications Office of the European Union.

European Commission, Directorate-General for Research and Innovation. (2023). *Integration of social sciences and humanities in Horizon 2020: Participants, budgets and disciplines 2014–2020—Final monitoring report*. Luxembourg: Publications Office of the European Union.

European Union. (2021). *Regulation (EU) 2021/1119 of the European Parliament and of the Council of 30 June 2021 establishing the framework for achieving climate neutrality and amending Regulations (EC) No 401/2009 and (EU) 2018/1999 ('European Climate Law')*. Brussels: Official Journal of the European Union.

European Union. (2023a). *Directive (EU) 2023/1791 of the European Parliament and of the Council of 13 September 2023 on energy efficiency and amending regulation (EU) 2023/955 (recast)*. Brussels: Official Journal of the European Union.

European Union. (2023b). *Directive (EU) 2023/2413 of the European Parliament and of the Council of 18 October 2023 amending Directive (EU) 2018/2001, Regulation (EU) 2018/1999 and Directive 98/70/EC as regards the promotion of energy from renewable sources, and repealing Council Directive (EU) 2015/652*. Brussels: Official Journal of the European Union.

European Union. (2024). *Directive (EU) 2024/1275 of the European Parliament and of the Council of 24 April 2024 on the energy performance of buildings (recast)*. Brussels: Official Journal of the European Union.

Foulds, C., & Robison, R. (2018). Mobilising the energy-related social sciences and humanities. In C. Foulds & R. Robison (Eds.), *Advancing energy policy* (pp. 1–11). Palgrave Macmillan.
Foulds, C., Royston, S., Berker, T., Nakopoulou, E., Bharucha, Z. P., Robison, R., Abram, S., Ančić, B., Arapostathis, S., Badescu, G., Bull, R., Cohen, J., Dunlop, T., Dunphy, N., Dupont C., Fischer, C., Gram-Hanssen, K., Grandclément, C., Heiskanen, E., Labanca, N., Jeliazkova, M., Jörgens, H., Keller, M., Kern, F., Lombardi, P., Mourik, R., Ornetzeder, M., Pearson, P., Rohracher, H., Sahakian, M., Sari, R., Standal, K., & Živčič, L. (2022). An agenda for future Social Sciences and Humanities research on energy efficiency: 100 priority research questions. *Humanities and Social Sciences Communications, 9*(1), 1–18.
Ingeborgrud, L., Heidenreich, S., Ryghaug, M., Skjølsvold, T. M., Foulds, C., Robison, R., Buchmann, K., & Mourik, R. (2020). Expanding the scope and implications of energy research: A guide to key themes and concepts from the social sciences and humanities. *Energy Research & Social Science, 63*, 101398. https://doi.org/10.1016/j.erss.2019.101398
Kropp, K. (2021). The EU and the social sciences: A fragile relationship. *The Sociological Review, 69*(6), 1325–1341. https://doi.org/10.1177/003802 61211034706
Krupnik, S., Wagner, A., Vincent, O., Rudek, T. J., Wade, R., Mišík, M., Akerboom, S., Foulds, C., Smith Stegen, K., Adem, Ç, Batel, S., Rabitz, F., Certomà, C., Chodkowska-Miszczuk, J., Denac, M, Dokupilová, D., Leiren, M. D., Frolova Ignatieva, M., Gabaldón-Estevan, D., Horta, A., Karnøe, P., Lilliestam, J., Loorbach, D., Mühlemeier, S., Nemoz, S., Nilsson, M., Osička, J., Papamikrouli, L., Pellizioni, L., Sareen, S., Sarrica., M., Seyfang, G., Sovacool, B., Telešienė, A., Zapletalová, V., & von Wirth, T. (2022). Beyond technology: A research agenda for social sciences and humanities research on renewable energy in Europe. *Energy Research & Social Science, 89*, 102536. https://doi.org/10.1016/j.erss.2022.102536
Robison, R., Skjølsvold, T. M., Hargreaves, T., Renström, S., Wolsink, M., Judson, E., Pechancová, V., Demirbağ-Kaplan, M., March, H., Lehne, J., Foulds, C., Bharucha, Z., Bilous, L., Büscher, C., Carrus, G., Darby, S., Douzou, S., Drevenšek, M., Frantál, B., Guimarães Pereira, Â., & Wyckmans, A. (2023). Shifts in the smart research agenda? 100 priority questions to accelerate sustainable energy futures. *Journal of cleaner production, 419*, 137946. https://doi.org/10.1016/j.jclepro.2023.137946
Royston, S., & Foulds, C. (2021). The making of energy evidence: How exclusions of Social Sciences and Humanities are reproduced (and what researchers can do about it). *Energy Research & Social Science, 77*, 102084. https://doi.org/10.1016/j.erss.2021.102084

Ryghaug, M., Subotički, I., Smeds, E., von Wirth, T., Scherrer, A., Foulds, C., Robison, R., Bertolini, L., Beyazit İnce, E., Brand, R., Cohen-Blankshtain, G., Dijk, M., Freudendal Pedersen, M., Gössling, S., Guzik, R., Kivimaa, P., Klöckner, C., Lazarova Nikolova, H., Lis, A., Marquet, O., Milakis, D., Mladenović, M., Mom, G., Mullen, C., Ortar, N., Paola, P., Sales Oliveira, C., Schwanen, T., Tuvikene, T., & Wentland, A. (2023). A Social Sciences and Humanities research agenda for transport and mobility in Europe: Key themes and 100 research questions. *Transport Reviews, 43*(4), 755–779. https://doi.org/10.1080/01441647.2023.2167887

Silvast, A., & Foulds, C. (2022). *Sociology of interdisciplinarity: Dynamics of energy research.* Palgrave Macmillan.

Open Access This chapter is licensed under the terms of the Creative Commons Attribution 4.0 International License (http://creativecommons.org/licenses/by/4.0/), which permits use, sharing, adaptation, distribution and reproduction in any medium or format, as long as you give appropriate credit to the original author(s) and the source, provide a link to the Creative Commons license and indicate if changes were made.

The images or other third party material in this chapter are included in the chapter's Creative Commons license, unless indicated otherwise in a credit line to the material. If material is not included in the chapter's Creative Commons license and your intended use is not permitted by statutory regulation or exceeds the permitted use, you will need to obtain permission directly from the copyright holder.

PART II

Navigating Community Participation

CHAPTER 2

Simplify the Uptake of Community Energy by Leveraging Intermediaries and the Use of Digital Planning Tools

Franziska Mey, Kristian Borch, Stephan Bosch, Benita Ebersbach, Robert Hecht, Lars Holstenkamp, and Jörg Radtke

Policy Highlights To achieve the recommendation stated in the chapter title, we propose the following:

- Ensure the follow-up and implementation of EU policy measures, including by conducting quality assessments and introducing national community energy targets.
- Foster institutional allies at local and regional levels.

F. Mey (✉) · B. Ebersbach · J. Radtke
Research Institute for Sustainability, Helmholtz Centre Potsdam, Potsdam, Germany
e-mail: franziska.mey@rifs-potsdam.de

B. Ebersbach
e-mail: benita.ebersbach@rifs-potsdam.de

J. Radtke
e-mail: joerg.radtke@rifs-potsdam.de

© The Author(s) 2024
A. Crowther et al. (eds.), *Strengthening European Energy Policy*,
https://doi.org/10.1007/978-3-031-66481-6_2

- Provide access to and capacities for using digital planning tools.
- Nurture knowledge and method integration across STEM and SSH disciplines to develop practices and tools to effectively implement community-led initiatives.

Keywords Energy transition · Citizen participation · Local governance

2.1 Introduction

There is growing consensus that community-led renewable energy initiatives play a crucial role for energy transitions delivering on both net-zero and just transition objectives (Bauwens et al., 2022; Standal et al., 2023). While moving away from fossil fuels demands collective efforts from a diverse array of stakeholders at different technology scales, these decentralised and bottom-up social innovations are commonly cited for ensuring greater citizen participation (Bielig et al., 2022; Creamer et al.,

---

K. Borch
Aalborg University, Copenhagen, Denmark

Ruralis, Trondheim, Norway

K. Borch
e-mail: kristian.borch@ruralis.no

S. Bosch
Institute of Geography, University of Augsburg, Augsburg, Germany
e-mail: stephan.bosch@geo.uni-augsburg.de

R. Hecht
Leibniz Institute of Ecological Urban and Regional Development (IOER), Dresden, Germany
e-mail: r.hecht@ioer.de

L. Holstenkamp
ECOLOG-Institute of Social-Ecological Research and Education (Non-profit), Leuphana University, Lüneburg, Germany
e-mail: lars.holstenkamp@ecolog-institut.de

2019). This is not only crucial for ensuring public support and acceptance of renewable energy projects, it also helps to promote sustainable energy intentions and behaviours, creating social norms that accelerate the energy transition locally (Sloot et al., 2018). Indeed, the last decade has seen a growing number of communities and citizen initiatives taking a more active role in the decarbonisation of the energy system, in terms of generation, consumption, and distribution, across both rural and urban regions of Europe. Currently, the European Federation of Citizen Energy Cooperatives has a network of 2250 initiatives operating across Europe, jointly representing over 1.5 million citizens (REScoop, 2024).

Community-driven solutions have a great appeal as they can add further renewable generation capacity into the energy mix, with this increasing the flexibility of the energy system, and diversifying the actor base and decision-making authority beyond the traditional centralised incumbents. In fact, non-technical benefits dominate the current narrative, which outlines the positive impacts and abilities of such initiates to integrate citizens' needs and opinions, mitigate resistance against transition measures, provide fairer models of prosumership, and strengthen democratic control and energy justice.

The narratives of the social impact have particularly shaped the conceptualisation of community-led energy initiatives and their perception for instigating a new relationship between society and its energy system. Hence policymakers from the European Union (EU) have acknowledged the importance of providing a legal framework with two variants of community energy actions—'Citizen Energy Communities' (CEC) and 'Renewable Energy Communities' (REC)—as further explained in Table 2.1, which details the broader understanding of the concept in the literature across purpose, governance, and activity types (European Union, 2023).

Although these acknowledgements are notable, and researchers, as well as policymakers, are increasingly well-versed in what community energy should mean (Creamer et al., 2019), challenges do persist when it comes to implementation, practice, and impact. In fact, the accelerated pace of the energy transition increases the risk of perpetuating existing structural inequalities and falling short on achieving a democratic just transition. Consequently, much still needs to be done to catalyse growth in the community energy sector and to leverage positive social benefits. Yet, we acknowledge that the transition to a decentralised renewable energy supply does not inherently guarantee a fairer, more gender-equitable,

Table 2.1 Definitions of community-led energy initiatives by the European Commission (Inspired by European Union, 2023)

Terms	Citizen Energy Community (CEC)	Renewable Energy Community (REC)
Governance and members	A legal entity that is based on voluntary and open participation, effectively controlled by shareholders or members who are natural persons, local authorities, including municipalities, or small enterprises, and micro-enterprises	A legal entity that, in accordance with the applicable national law, is based on open and voluntary participation, autonomous, effectively controlled by shareholders or members that are located in the proximity of the renewable energy projects that are owned and developed by that legal entity; the shareholders or members of which are natural persons, SMEs, or local authorities, including municipalities
Purpose	Both share the purpose to provide environmental, economic, or social community benefits for its shareholders, members or the local areas where they operate, rather than financial profit	

or more inclusive system. Socio-demographic and structural inequalities (including those related to income, education, and health) will persist, even in greener and more ecological systems, unless politics, business, and society actively address these disparities and prioritise the needs and preferences of the public. We posit that community-led initiatives offer the potential to address these issues.

In this chapter, we draw upon interdisciplinary insights across Geographic Science, Political Science, Psychology, and Economics, and incorporate valuable feedback from practitioners and advocates. We therefore bridge across the Social Sciences and Humanities (SSH) and Science, Technology, Engineering, and Mathematics (STEM) disciplines to emphasise the political contexts, governance factors, and spatial analytical techniques necessary to foster community-led initiatives. The insights were gathered in an expert workshop in late-2023 (Mey et al., 2024a), which served to align our understanding of the subject, as well as brainstorm and prioritise recommendations, while building on data and knowledge gained across a broad range of research projects, including:

- Studies analysing the socio-economic dimensions of community energy projects, in particular their motivation, participatory practices, governance structures, and impacts (Holstenkamp & Kahla, 2016; Mey et al., 2022; Radtke, 2014).
- Research projects to understand the role of individual community energy actors as prosumers in the energy transition (EU Horizon 2020 project, PROSEU).
- Analysis of regional added value and financial benefits of community-led initiatives to support structural change processes (Energy Communities: Structural Policy & Participation [BE:ST]).
- Research linking to the spatial dimensions of the energy transition and the emphasis of participatory arrangements as crucial for establishing landscape democracy (CIVIC Renewables).
- Geographic Information Systems (GIS) technology-based research, highlighting an actor-centred approaches in spatial optimisation and energy modelling.
- Global initiatives like the Colouring Cities Research Programme and the Colouring Cities open data platforms (Hecht et al., 2023), gaining insights how to utilise web-based tools in a citizen's science approach for energy data gathering.

The initial and revised versions of this chapter were circulated among practitioners, as part of an open feedback process in early 2024, with their input used to refine the subsequent text.

Our aim is to offer energy policy recommendations to the EU and its Member States, to enhance the resilience of community-led energy initiatives, and to provide suggestions on: how to fully unlock their potential by leveraging various instruments at different governmental levels; and, access to information resources.

2.2 Strengthen Community-Led Renewable Energy Initiatives

2.2.1 *Ensure Follow-Up and Implementation of EU Policy Measures*

Community energy policies need diligent follow-through, including instruments for quality assessment. Since the introduction of energy communities into EU law with the adoption of the 'Clean energy for all Europeans' legislative package in 2019 (European Commission, 2019), the EU has started several initiatives to strengthen collective initiatives in the energy sector. These include policy proposals as part of the Solar Energy Strategy (European Commission, 2022), as well as research and technical assistance in the context of the Energy Communities Repository and the Rural Energy Communities Advisory Hub (European Commission, 2023, 2024). Yet, implementation requires genuine follow-through; the EU must hold Member States accountable for assessing the status of EU policies for community energy and the evaluation of measures taken. Here, the introduction of a national target for energy communities could contribute clear direction, show commitment, and guide decision-making at various governmental levels.

According to the recast Renewable Energy Directive (RED II) (European Union, 2018), Member States shall take out assessments of the community energy sector in their countries, specifically of barriers and potentials (RED II, Art. 22 Para. 3). These assessments should then build the basis of an enabling framework (RED II, Art. 22 Para. 4), which Member States should set up. In addition to this, Member States could also use regular assessments to monitor the efficiency and effectiveness of policy measures. Although the EU legislator has not (yet) defined any standards nor provided any template for such an assessment, there are already proposals that exist (Holstenkamp & Kriel, 2022).

An assessment needs to build on data about the current state and the development of energy communities, ideally disaggregated for different types of energy communities. In this regard, Member States can partly build on data provided by national umbrella organisations, through the Energy Community Repository and/or by European research consortia. Moreover, as is well known from the evaluation literature, an assessment should build on an impact model or a theory of change (Weiss, 1995). As known from the Management literature (Drucker, 1973), *sensible* measurement that provides feedback into the management process is important—meaning that the process needs to involve the build-up of relationships (from participation in the energy community sector) to make sense of the data and manage them in a correct way.

According to the EU Solar Energy Strategy (which is part of the REPowerEU plan), EU Member States shall ensure to set up at least one renewables-based energy community in every municipality with a population above 10,000 by 2025 (European Commission, 2022). However, almost all Member States are yet to implement any concrete national targets for energy communities. Such targets make it easier to follow-up and report on developments in the sector, e.g. in the National Energy and Climate Plans (NECPs). This is why the European Commission stresses the lack of quantitative targets for energy communities in nearly all its assessments of NECPs. The structure and format of NECPs laid out in the Implementing Regulation foresees targets for energy communities.

Clear targets or goals are supposed to create a management tool that "help[…] to ensure the accountability of all stakeholders" (SDSN, 2015, p. 2). They could direct (political) attention towards energy communities. What needs to be considered, though, is that indicators also have political dimensions—again highlighting the importance of embedding the selection and use of targets into a participatory process.

2.2.2 Foster Institutional Allies at Local and Regional Levels

Since community-led energy initiatives originate at the grassroots-level, distinct from commercial entities, they often lack the knowledge as well as the capacity to plan and implement energy projects. Hence studies have shown that community-led renewable energy initiatives can: greatly benefit from intermediary organisations enabling learning, skills, and knowledge transfers; provide capacities and resources; and aggregate impact for stronger local energy activities (Arler et al., 2023; Mey et al., 2016). This rationale gave impetus for the establishment of one-stop shops at EU (e.g. Energy Communities Repository) and National levels (e.g. Spain, Netherlands, France, Belgium, Croatia). An example is the Coordination Office for Energy Communities in Austria, with the role to provide practical advice, knowledge, materials, and resources for new and operating energy communities (Österreichische Koordinationsstelle für Energiegemeinschaften, 2024). In addition, this institution has also created regional and local subsidiaries offering personal consultation sessions.

While we acknowledge positive strides are being made in some countries, greater action is necessary to support communities at regional and local levels across most countries in the EU. The last decade has seen an increasing engagement of particularly local governments in the fledgling community energy sector, with many local climate and energy initiatives emerging across the EU. However, these institutional partnerships are often underfunded and have lower capacities to navigate the complexity of the energy-related legal frameworks, regulatory instruments, and procedures. We found that these limitations may increase a reluctance to act and hinder the brokering role of local government to engage in partnerships and collaborations, which in turn may hamper communication, community cohesion, and trust.

Hence, we argue that while local governments already play a significant role in climate action, they often lack the capacities and resources to take on additional tasks or fulfil their existing responsibilities to a satisfactory extent. Policies—such as the amendment of the German Renewable Energy Act 2023 (Section 6 of the Act), which allows wind energy and open-space photovoltaic projects to financially involve affected local municipalities in the profits generated—are important developments and may give these institutions greater scope for action. Consequently, strengthening local governments and other intermediaries to better facilitate community participation in energy planning and energy project implementation is a crucial step to foster a decentralised, community accepted, and co-owned energy transition.

2.2.3 Provide Access and Increase Capacities for Using New Planning Tools

A growing field of conflict is the selection of sites for renewable energy projects because legitimate concerns from affected communities are often neglected in the national legal framings (Borch et al., 2023). These unmet concerns may spin siting disagreements out-of-control as they often travel (overflow) to other settings (e.g. social media), escalating into entrenched conflicts (Borch et al., 2020). Here, it is crucial to optimise communication processes, providing comprehensive knowledge about the techno-economic and socio-ecological conditions of selected sites, and creating transparency regarding regulatory factors to communities.

In the last decade, innovative, user-oriented, web-based planning, and visualisation tools like GIS have received increasing recognition in planning and implementing renewable energy projects at local level (Bosch & Schwarz, 2018). These tools can enable participatory mapping of sites, political prioritisation of land use, and empower local communities to navigate competing land use interests.

We argue that community-led initiatives could benefit from utilising these tools to further strengthen their abilities for an inclusive and locally informed decision-making process for site selection. However, we find that these initiatives often face challenges due to limited resources, capacities, and information for accessing these tools.

In Germany, a growing movement is advocating for greater transparency and participation in local energy planning. An interesting example of this is the *Bavarian Energy Atlas*, which is provided by the *Bavarian Ministry of Economic Affairs, Regional Development and Energy*. While the digital atlas does not allow for complex spatio-temporal analyses of the implementation of renewable technologies, it does nevertheless provide valuable insights into the site potential and planning basis for renewable energies in the state of Bavaria. In order to get in touch with local actors and facilitate communication with local experts, important contacts are also mapped in the atlas. Moreover, users can select individual municipalities and calculate local energy potentials. Other participatory planning tools are also linked to the *Energy Atlas*, such as the 3D-analyser, which can be used to place wind or photovoltaic plants in the landscape and thus provide a visual-aesthetic impression of the specific spatial impact of renewable energies.

Since these tools make it possible to simulate social, ecological, and economic environments, we argue that they can highlight trade-offs and help to make informed decisions at community level. GIS tools assist in simplifying the complexity of local planning processes and empowering local actors to actively engage in dialogue aimed at transforming the energy system. This is particularly achieved through the identification and prioritisation of suitable locations, and ideally enables a democratic dialogue, coordination, and balancing of the needs and preferences of the local population and stakeholders (Mey et al., 2024b).

However, to contribute to the acceleration of a local energy transition, a coordinated development of geodata infrastructures and a better access of geodata is necessary. Only then will it be possible to establish participatory and collaborative digital planning approaches, in contrast to the top-down mechanisms that still dominate energy planning processes at present. As some countries have already advanced in their transition, they can provide insights in appropriate geodata usage. For example, GIS and energy system analyses have been used to provide technical and socio-economic knowledge regarding the potential role of photovoltaic in an urban system, including the societal perspective in Denmark (Mathiesen et al., 2017).

Hence, optimising the use and accessibility of these tools can help accelerate the development of energy community projects. EU policy should encourage Member States to prioritise collaboration, adaptation, capacity building, and community engagement, to ensure that existing tools become an integral part of communities' journey towards sustainable energy practices.

2.3 Achieving Our Recommendation

We recognise the increasing importance of community-led energy initiatives in achieving both net-zero and just transition objectives. Although, in the recent decades, community-led initiatives have enjoyed the support of various policy measures, we find that continuous efforts from EU and Member State level are needed to facilitate and simplify their implementation.

As per the title of this chapter, our core recommendation is that policy should: *simplify the uptake of community energy by leveraging intermediaries and the use of digital planning tools*. We argue that this can be achieved through the following sub-recommendations: firstly, it is important to strengthen overall community energy policies in the Member States. To this purpose, it is necessary to ensure a follow-up and implementation of EU RED II and its CEC and REC provisions, further enhanced by quality assessments of national policy measures and the introduction of a national community energy target to evaluate the progress in the sector. Secondly, institutional allies at multiple levels are important factors for the success of local energy initiatives, yet still often lack resources and capacities to do so. Hence, empowering local governments and intermediaries to enhance community involvement in energy planning and project implementation is essential for promoting a decentralised and co-owned energy transition. In particular, we consider the new digital planning tools as great opportunities to simplify processes, empower local actors, and speed up energy community project development. Therefore, our third recommendation is to enhance tool accessibility and user capacity at the local level. Finally, we conclude that integrating insights, tools, and methods from both SSH and STEM disciplines is essential to leverage on social-political and technical opportunities for an effective implementation of community-led initiatives.

Acknowledgements We would like to thank the two reviewers for their insightful contributions, which greatly enhanced the quality of this chapter. We also appreciate the feedback provided by practitioners and advocates, whose valuable insights further enriched our work.

References

Arler, F., Sperling, K., & Borch, K. (2023). Landscape democracy and the implementation of renewable energy facilities. *Energies*, 16(13). https://doi.org/10.3390/en16134997

Bauwens, T., Schraven, D., Drewing, E., Radtke, J., Holstenkamp, L., Gotchev, B., & Yildiz, Ö. (2022). Conceptualizing community in energy systems: A systematic review of 183 definitions. *Renewable and Sustainable Energy Reviews*, 156. https://doi.org/10.1016/j.rser.2021.111999

Bielig, M., Kacperski, C., Kutzner, F., & Klingert, S. (2022). Evidence behind the narrative: Critically reviewing the social impact of energy communities in Europe. *Energy Research and Social Science*, 94, 102859. https://doi.org/10.1016/j.erss.2022.102859

Borch, K., Kirkegaard, J. K., & Nyborg, S. (2023). Three wind farm developments, three different planning difficulties: Cases from Denmark. *Energies*, 16(4662), 1–14. https://doi.org/10.3390/en16124662

Borch, K., Munk, A. K., & Dahlgaard, V. (2020). Mapping wind-power controversies on social media: Facebook as a powerful mobilizer of local resistance. *Energy Policy*, 138, 111223. https://doi.org/10.1016/j.enpol.2019.111223

Bosch, S., & Schwarz, L. (2018). Ein GIS-Planungstool für erneuerbare Energien – Integration sozialer Perspektiven, 92–101. https://doi.org/10.14627/537647012.Dieser

Creamer, E., Taylor Aiken, G., van Veelen, B., Walker, G., & Devine-Wright, P. (2019). Community renewable energy: What does it do? Walker and Devine-Wright (2008) ten years on. *Energy Research & Social Science*, 57, 101223. https://doi.org/10.1016/j.erss.2019.101223

Drucker, P. F. (1973). *Management : Tasks, responsibilities, practices. TA - TT -.* New York SE - XVI, 839 Seiten; 24 cm: Harper & Row New York. https://worldcat.org/title/729131336

European Commission. (2019). *Clean energy for all Europeans*. Luxembourg: Publications Office of the European Union.

European Commission. (2022). *EU solar energy strategy*. COM(2022) 221 final. Brussels: European Commission.

European Commission. (2023). *Rural Energy Community Hub*. https://rural-energy-community-hub.ec.europa.eu/index_en

European Commission. (2024). *Energy communities repository—One-stop shops*. https://energy-communities-repository.ec.europa.eu/energy-communities-repository-support/energy-communities-repository-one-stop-shops_en

European Union. (2018). *Directive (EU) 2018/2001 of the European Parliament and of the Council of 11 December 2018 on the promotion of the use of energy from renewable sources (recast)*. Brussels: Official Journal of the European Union.

European Union. (2023). *Directive (EU) 2023/2413 of the European Parliament and of the Council of 18 October 2023 amending Directive (EU) 2018/2001, Regulation (EU) 2018/1999 and Directive 98/70/EC as regards the promotion of energy from renewable sources, and repealing Council Directive EU 2015/652*. Brussels: Official Journal of the European Union.

Hecht, R., Danke, T., Herold, H., Hudson, P., Munke, M., & Rieche, T. (2023). Colouring Cities: A Citizen Science Platform for Knowledge Production on the Building Stock - Potentials for Urban and Architectural History. In S. Münster, A. Pattee, C. Kröber, & F. Niebling (Eds.), *Research and education in urban history in the age of digital libraries. UHDL 2023. Communications in computer and information science* (Vol. 1853). Springer. https://doi.org/10.1007/978-3-031-38871-2_9

Holstenkamp, L., & Kahla, F. (2016). What are community energy companies trying to accomplish? An empirical investigation of investment motives in the German case. *Energy Policy, 97*, 112–122. https://doi.org/10.1016/j.enpol.2016.07.010

Holstenkamp, L., & Kriel, C. (2022). Model assessment structure proposal. *Zenodo.* https://doi.org/10.5281/zenodo.7243930

Mathiesen, B. V., David, A., Petersen, S., Sperling, K., Hansen, K., Nielsen, S., Lund, H., & Neves, J. B. D. (2017). *The role of Photovoltaics towards 100% Renewable energy systems: Based on international market developments and Danish analysis.* Department of Development and Planning, Aalborg University. https://vbn.aau.dk/ws/portalfiles/portal/266332758/Main_Re[1]port_The_role_of_Photovoltaics_towards_100_percent_Renewable_Energy_Systems.pdf

Mey, F., Borch, K., Bosch, S., Ebersbach, B., Hecht, R., Holstenkamp, L., & Radtke, J. (2024a). Approach and method to developing the book chapter: Simplify the uptake of community energy. *Zenodo.* https://doi.org/10.5281/zenodo.11160172

Mey, F., Diesendorf, M., & MacGill, I. (2016). Can local government play a greater role for community renewable energy? A case study from Australia. *Energy Research & Social Science, 21*, 33–43. https://doi.org/10.1016/j.erss.2016.06.019

Mey, F., Kallies, A., Wiseman, J., & Watson, M. (2022). Legitimizing energy transitions through community participation: Germany and Australia at a crossroad. *Globalizations*, 1–19. https://doi.org/10.1080/14747731.2022.2138261

Mey, F., Lilliestam, J., Wolf, I., & Tröndle, T. (2024b). Visions for our future regional electricity system: Citizen preferences in four EU countries. *IScience*. https://doi.org/10.1016/j.isci.2024.109269

Österreichische Koordinationsstelle für Energiegemeinschaften. (2024). *Informationsplattform Energiegemeinschaften.* https://energiegemeinschaften.gv.at/

Radtke, J. (2014). A closer look inside collaborative action: Civic engagement and participation in community energy initiatives. *People, Place and Policy Online, 8*(3), 235–248. https://doi.org/10.3351/ppp.0008.0003.0008

REScoop. (2024). *REScoop.* https://www.rescoop.eu/

SDSN. (2015). Indicators and a monitoring framework for the Sustainable Development Goals: Launching a data revolution for the SDGs. *A Report by the Leadership Council of the Sustainable Development Solutions Network,* 160.

Sloot, D., Jans, L., & Steg, L. (2018). Can community energy initiatives motivate sustainable energy behaviours? The role of initiative involvement and personal pro-environmental motivation. *Journal of Environmental Psychology, 57,* 99–106. https://doi.org/10.1016/j.jenvp.2018.06.007

Standal, K., Leiren, M. D., Alonso, I., Azevedo, I., Kudrenickis, I., Maleki-Dizaji, P., Laes, E., Di Nucci, M., & Krug, M. (2023). Can renewable energy communities enable a just energy transition? Exploring alignment between stakeholder motivations and needs and EU policy in Latvia, Norway, Portugal and Spain. *Energy Research and Social Science, 106,* 103326. https://doi.org/10.1016/j.erss.2023.103326

Weiss, C. H. (1995). Nothing as practical as good theory. In *New Approaches to Evaluating Community Initiatives* (pp. 65–92).

Open Access This chapter is licensed under the terms of the Creative Commons Attribution 4.0 International License (http://creativecommons.org/licenses/by/4.0/), which permits use, sharing, adaptation, distribution and reproduction in any medium or format, as long as you give appropriate credit to the original author(s) and the source, provide a link to the Creative Commons license and indicate if changes were made.

The images or other third party material in this chapter are included in the chapter's Creative Commons license, unless indicated otherwise in a credit line to the material. If material is not included in the chapter's Creative Commons license and your intended use is not permitted by statutory regulation or exceeds the permitted use, you will need to obtain permission directly from the copyright holder.

CHAPTER 3

Prioritise Inclusive, Early, and Continuous Societal Engagement to Maximise the Benefits of Geothermal Technologies

Melanie Rohse, Amel Barich, Claire Bossennec, Annick Loschetter, Adele Manzella, Anna Pellizzone, Stacia Ryder, and Iain Soutar

Policy Highlights To achieve the recommendation stated in the chapter title, we propose the following:

- Policymakers should mandate that energy decision-making processes be more inclusive, supporting both geothermal representation in large-scale energy deliberation and the inclusive engagement of local communities in individual geothermal projects.
- Policymakers should foster inclusive societal engagement that goes beyond the minimal requirements of consulting and informing at

M. Rohse (✉)
Global Sustainability Institute, Anglia Ruskin University, Cambridge, UK
e-mail: melanie.rohse@aru.ac.uk

A. Barich
GEORG—Geothermal Research Cluster, Reykjavík, Iceland
e-mail: amel@georg.cluster.is

© The Author(s) 2024
A. Crowther et al. (eds.), *Strengthening European Energy Policy*,
https://doi.org/10.1007/978-3-031-66481-6_3

planning stage, to incorporate processes of deliberation and co-creation with local communities.
• Geothermal industry actors should practise inclusive societal engagement which is early, continuous, and sensitive to the technical specificities (e.g. local resource, subsurface uncertainties) and social challenges (e.g. low public awareness) of geothermal technologies.
• Language alignment activities and mutual expert elicitation can support the exchange of knowledge across social-technical disciplines, bridging disciplinary siloes to tackle the problem above.

Keywords Social acceptance · Renewable heat · Place-based deliberation · Multi-scale participation · Energy justice

C. Bossennec
Section 4.8: Geoenergy, GFZ German Research Centre for Geosciences, Potsdam, Germany
e-mail: claire.bossennec@gfz-potsdam.de

A. Loschetter
Bureau de Recherches Géologiques et Minières (BRGM), Orléans, France
e-mail: a.loschetter@brgm.fr

A. Manzella
Institute of Geosciences and Earth Resources, Consiglio Nazionale Delle Ricerche, Pisa, Italy
e-mail: adele.manzella@cnr.it

A. Pellizzone
Milan, Italy

S. Ryder
Department of Sociology and Anthropology, Utah State University, Logan, UT, USA
e-mail: stacia.ryder@usu.edu

I. Soutar
Centre for Geography and Environmental Science, University of Exeter, Penryn, UK
e-mail: i.soutar@exeter.ac.uk

3.1 Introduction

A key ambition of the European Green Deal is to supply "clean, affordable and secure energy" (European Commission, 2019, p. 6), which has been bolstered by measures to rapidly diversify and roll-out renewable energy technologies in the REPowerEU plan[1] (European Commission, 2022). In this context, geothermal energy has a major role to play in moving the EU away from our reliance on gas, as recognised in the Revised Renewable Energy Directive in 2023 (European Union, 2023), yet it is not currently realising its potential across Europe. At the same time, the European Green Deal aspires to a "just and inclusive" transition (European Commission, 2019, p. 2) where communities and citizens can work in partnership with institutions and organisations in energy decision-making. However, societal engagement tends to be neglected in strategic documents on upscaling geothermal technologies, e.g. the recent Implementation Plan of the Geothermal Implementation Working Group only mentions communities in passing (Geothermal IWG, 2023), and inadequate engagement risks eroding the social legitimacy of geothermal energy and may limit the potential to achieve just transition ambitions.

To address the challenge of increasing the uptake of geothermal energy whilst including communities in energy decision-making, we assembled an interdisciplinary team of international experts, representing several areas of Geoscience (including Geophysics, Subsurface Exploration, and Predictive Modelling); Science Communication; Citizen Engagement with Research and Innovation; Environmental Social Science; Social Geothermal Sciences; Human Geography; and Environmental Sociology.

Over the course of four months, we engaged in iterative mutual expert elicitation (i.e. gathering expert knowledge and judgement) through several preparatory virtual meetings, followed by a face-to-face, in-depth interactive workshop to tackle three key themes: (1) understanding different types of geothermal technologies and their associated risks and benefits, (2) how to advance the role of society in transition to climate neutrality, and (3) how to increase engagement at a project level, given the specificities of geothermal projects. In our face-to-face workshop, we worked through disciplinary definitions of key terms and exchanged knowledge of state-of-the-art research—materials which were

---

[1] This EU energy plan launched in 2022, in part as an attempt to make Europe independent from Russian energy following the Russian invasion of Ukraine.

co-produced at the workshop, including mind maps, are available open access (Rohse et al., 2024). Our mutual expert elicitation enabled us to understand disciplinary knowledges and how disciplinary siloes may be bridged.

At the Geoscience end of the spectrum, we explored the complexity of geothermal technologies, their specific risks and mitigation measures, and discussed how the initial set-up can be costly and lengthy. We also observed how the Social Sciences tend to be neglected within geothermal projects, uncovering how top-down approaches to engagement (e.g. those focused on information sharing) tend to dominate. From the Social Science perspective, we unpacked many different understandings of engagement, and discussed examples of good practice, both from within and outside the geothermal sector. Our key learning concerns the abstract nature of Social Science frameworks and the need to translate them into practice. Our interdisciplinary encounter therefore allowed us to consider those abstract frameworks in the 'real-world' context of several geothermal technologies.

As a result, we provide recommendations on incorporating societal engagement throughout geothermal project developments aimed at: (1) decision-makers at EU and national levels, as they set national energy agendas and should be responsible for giving societal engagement an equitable place in policymaking, (2) local and regional decision-makers, as they plan and implement local energy plans, and 3) geothermal operators across Europe, as they propose and implement projects.

In the main body of the chapter, we outline (1) current understanding of geothermal technologies, (2) how geothermal energy may be introduced in EU, national and local energy decision-making, and (3) how societal engagement in geothermal projects can become more inclusive.

3.2   Technology and Society in the Geothermal Sector

3.2.1   Understanding Geothermal Technologies

One goal of our interdisciplinary encounter was to ensure that we shared disciplinary perspectives on geothermal technologies, including their diversity and their associated opportunities, risks, and challenges. This enabled us to consider how these specificities may play out in practices of societal engagement, as we explore below.

Geothermal energy can be found everywhere. It is a non-intermittent flexible resource, which offers promising opportunities in the renewable energy landscape. Geothermal technologies provide a sustainable energy source, with generally minimal greenhouse gas emissions and a light environmental footprint overall. In terms of societal perceptions, we hypothesise that the low-carbon nature of the technology could make it appealing, bearing in mind that opposition to another subsurface technology, hydraulic fracturing for shale gas exploitation, has been partly driven by concerns about the impacts of fossil fuels on the climate. Indeed, in our experience in the UK, geothermal energy has appeared to have a receptive welcome, with community members near some UK geothermal sites reflecting on how geothermal energy can help shift us away from a reliance on fossil fuels. However, we acknowledge this is only part of the picture. Whilst, as a subsurface activity, geothermal energy is almost invisible and takes up a relatively low surface area, there are impacts related to surface operations (e.g. land occupation, visual impact, noise) that could affect local communities, and raise concerns over the distribution of costs and benefits of these developments.

There is a diversity of applications for geothermal energy, which can be harnessed for electricity generation, heating and cooling, hot water, and minerals supply (e.g. lithium). This diversity of applications means that the range of risks and impacts are broad and vary. However, the environmental risks and technical challenges associated with geothermal developments are well-documented and mitigated thanks to regulation and best practices (Chen et al., 2020; Gombert et al., 2018). For example, geo-mechanical changes (e.g. seismicity) and underground changes (e.g. disturbance of non-targeted aquifers) both require careful monitoring and mitigation. In terms of societal engagement, some of our research in the UK shows seismicity was a concern in other subsurface sectors (Ryder et al., 2023). Yet, seismicity (and other environmental risks, like chemical leakages) is not a risk for all geothermal applications. Therefore, we highlight the importance of developers being clear and transparent about risks and mitigation for proposed technologies as paramount for trust-building and engagement.

In addition, financial challenges exist. Initial drilling and exploration can be relatively costly and lengthy, which makes attracting investment challenging, especially as the uncertainties relating to variability of the underground and to gaps in knowledge tend to be high before drilling and can remain significant during operation. The length of time it can take for geothermal projects to be set up and completed can have implications for engagement. On the one hand, a longer process provides opportunity for more engagement; on the other hand, supporters may eventually lose faith in a project ever coming to fruition.

3.2.2 Introducing Geothermal Energy in EU, National and Local Energy Decision-Making

The growth of geothermal technology holds the potential to significantly contribute to a cleaner and more sustainable energy future. Yet, public awareness of geothermal technologies is low, and societal engagement in geothermal developments tends to be top-down and focused on acceptance. This is in contrast with EU efforts to make citizens' voices heard (e.g. via the Conference on the Future of Europe[2]), including specifically on the transition to climate neutrality. As a result, we propose that (1) policymakers should support initiatives to make energy decision-making more inclusive, ensuring that geothermal is appropriately represented in deliberation processes on energy futures at EU and national levels, and in deliberations about local technology choices, and (2) that operator- and developer-led engagement can be bottom-up and inclusive (see Sect. 3.2.3). We start by exploring the first of these below.

Public deliberation exercises[3] are proliferating in Europe at various scales—for example, national climate assemblies, such as in Denmark (The Citizens' Assembly on Climate Issues, 2021), and at regional level, such as in Lombardy, Italy (Simone et al., 2023), to collectively design the path towards carbon neutrality, and align public strategies and actions to the views, needs, and concerns of communities. This type of broad societal engagement can enable a more inclusive and just energy transition. For example, it can contribute to setting goals beyond electoral mandates, giving legitimacy to public authorities' choices, and building trust and alliances between different societal actors.

---

[2] Details available via: https://commission.europa.eu/strategy-and-policy/priorities-2019-2024/new-push-european-democracy/conference-future-europe_en.

[3] A detailed definition and description of public deliberation for policymaking is available here: https://www.oecd-ilibrary.org/sites/339306da-en/index.html?itemId=/content/publication/339306da-en.

We argue that EU and national policymakers should support public deliberation exercises on the future of energy in which geothermal technologies are presented within a range of potentially viable options. This is necessary due to the geothermal sector being at an earlier stage of expansion compared to other sectors. Indeed, there are practical challenges associated with representing a full range of energy technologies at an appropriate level of detail to communities (Elstub et al., 2021). As such, less common technologies such as geothermal ones may be at a disadvantage when compared to more mainstream options with which the public is more familiar, and where more evidence about effectiveness exists.

The relative novelty of geothermal technologies means in some areas, the potential economic, environmental, and system benefits may be unknown or unclear. Similarly, the novel nature of geothermal projects means that perceptions about associated risks may not be based in real-world experiences or may be affected by 'spillover' from other projects, where negative perception of other energy projects lead to negative perceptions of geothermal projects (Westlake et al., 2023). As such, there is a need for objective evidence to help guide decision-making processes, whilst also recognising and valuing knowledges and expertise from within communities, such as local underground knowledge (Ryder et al., 2023). Deliberative processes are ideal to bring those different types of knowledges together.

In addition, deliberative exercises have the potential to support communities in identifying appropriate geothermal opportunities at a smaller scale. Local authorities have a major role to play in this process. For instance, they can have identified geothermal energy as a *technically* viable option for local heating or cooling, and they need to work with local communities to establish whether it is a *socially* viable option, as in St. Gallen, Switzerland (Ejderyan et al., 2020). Within such processes, technical assessments must be transparent, emphasising a holistic view of potential risks and benefits across a project life cycle, and include different technologies to ensure that decision-making is informed and balanced. When this is enacted, communities can make informed decisions about future energy provision, including whether geothermal energy is 'right' for them and under what conditions, which deliberative and co-creative processes can help identify. Accepting the decisions that emerge from societal engagement, including if they reject a project or a technology, is essential.

Whilst there is a role for EU, national, and local policymakers in large-scale deliberation on the future of energy, or small-scale deliberation on local energy plans, geothermal projects are for the most part developer- and/or operator-led. In the next sub-section, we draw on our team's experiences with existing practices of engagement in geothermal energy, drawing attention to important considerations for developers and operators to widen their approaches to societal engagement.

3.2.3 How Societal Engagement in Geothermal Projects Can Become More Inclusive

As an author team, we collectively have in-depth experience working in several European contexts. In our workshop, we identified that the development of geothermal energy is inherently intertwined with local socio-economic contexts, and that societal engagement requires a case-specific approach. At the same time, through our international and interdisciplinary experience, we established the collective importance and desire for inclusive engagement to move the geothermal sector forward. Here, we provide examples of bottom-up and inclusive approach to societal engagement in the geothermal sector.

To our knowledge, a mark of success in geothermal developments is the ability to establish and sustain a high level of community trust and support. This achievement is directly tied to prioritising meaningful dialogue in the engagement process and leveraging mutual benefits. Fostering trust, which is closely connected to co-production and psychological identification with a project at the community level, is key in societal engagement and negotiating a "Social License to Operate" (Barich et al., 2022). A notable example is Iceland, where communities view geothermal projects not just as energy initiatives but as integral contributors to their well-being. Iceland has become a hub for knowledge transfer, capacity building and geothermal outreach. This has stemmed from a culture of open collaboration between sectors from within the country, which is being consciously maintained through meaningful dialogue with local communities and stakeholders.

How operators view, approach, and support communities, their willingness to be honest about risks and unknowns, and the way they accommodate and reduce community impacts are ways that industry-community relationships have been strengthened (Ryder et al. under review). One

UK example of this is when an operator chose to use a more expensive drill because of the noise reduction it would provide, reducing local disturbance.

Community-led geothermal initiatives, such as those in Madrid, Spain, (Hildebrand & Klein, 2022) and in Darmstadt,[4] Germany, are examples of residents coming together to shift their heating sources from fossil fuel to geothermal energy, reflecting a desire for greener technologies and heat supply. This demonstrates that under the right conditions, bottom-up projects are possible, and highlights the need for the industry to be visible and approachable to facilitate such participation in the energy system. Relatedly, alternative financial schemes such as crowdfunding for geothermal projects (e.g. Friederichs, 2021) can highly influence the deployment rate of geothermal energy.

However, there are some barriers for operators to gain community support. For instance, although several projects in the Upper Rhine Graben (e.g. Soultz-sous-Forets, France) proved their efficiency, poor project management and lack of transparency led to failure in other similar projects (e.g. Vendenheim, France), and strong rejection from local communities (Chavot et al., 2018), destroying trust in the sector in that area for years. Even in the positive examples of the UK, one operator who had previously engaged their community extensively faced pushback to a new project, in part due to a lack of in-depth engagement. This demonstrates the importance of early and continuous engagement within a project, and from one project to the next.

3.3 Achieving Our Recommendation

As per the title of this chapter, our core recommendation is that policy should *prioritise inclusive, early, and continuous societal engagement to maximise the benefits of geothermal technologies.*

Through our STEM-SSH collaboration, we have uncovered how the technical specificities of geothermal technologies and the social challenges they face intersect, and require interventions across scales from EU, national, local policymakers, and operators to ensure that the potential of geothermal energy in decarbonising heating, cooling, and electricity production can be realised, meeting the ambitions of the European Green

---

[4] See the DELTA project: https://delta-darmstadt.de/.

Deal and the REPowerEU plan for renewable energy technologies in a way that is fair to European communities.

Our iterative elicitation process enabled us to rapidly gain knowledge across disciplinary divides to identify points of connection between social and technical challenges, which led us to identify how several actors can support societal engagement specifically in geothermal developments, and recommend that policymakers and industry operators should prioritise inclusive, early, and continuous engagement across scales to maximise the benefits of geothermal technologies in the quest to supply clean energy across Europe (European Commission, 2019), whilst "promot[ing] the participation of local communities in renewable energy projects" (OJEU, 2023, p. 8).

At the local level, several examples in the geothermal sector demonstrate that valuing, resourcing, and implementing inclusive societal engagement is possible. Tools such as community-led initiatives, alternative modes of financing, deliberative energy decision-making and the Social License to Operate framework are ethical and transformational tools that can help broaden our understanding of societal engagement that can mutually benefit the geothermal sector and communities. However, these are scattered examples; embedding inclusive societal engagement requires formal support through EU, national, and local policy instruments in the process of scaling up geothermal technologies. To advance geothermal technologies as part of a fair energy transition, EU and national policymakers should mandate that energy decision-making processes be more inclusive, supporting both wider deliberation on the energy transition where the geothermal sector should be appropriately represented, and the inclusive engagement of local communities in geothermal projects at local and regional scales.

Local and regional policymakers should plan and implement inclusive societal engagement that goes beyond the minimal requirements of consulting and informing at planning stage, to incorporate processes of deliberation and co-creation with local communities. Whatever the mechanisms adopted, accountability of commissioning authorities is key. Anchoring engagement methods within public strategies or policies can help to ensure that public authorities act on co-created outcomes, making communities feel heard, which contributes to the credibility of such exercises.

Geothermal industry actors should practise inclusive societal engagement, which is early, continuous, and sensitive to the technical specificities (e.g. local resource; subsurface uncertainties) and social challenges (e.g. low public awareness) of geothermal technologies. Given the degree to which geothermal heating and cooling rely on local use, engagement with local communities is especially important for geothermal technologies. This local focus also opens doors for the possibility of more equitable distributions of the burdens and benefits of resource development. It is possible to expand the geothermal sector in a fair and inclusive way through: transparency about our understanding of the risks involved and of mitigation measures; designing projects addressing community members' expectations; maintaining meaningful dialogue; creating local benefits (e.g. local employment opportunities; domestic value chains); and establishing community-based projects. All of these require STEM-SSH collaborations.

Acknowledgements Work on this chapter was supported, in part, by PUSH-IT, funded by the European Union under the Horizon Europe programme (grant no. 1011096566). Views and opinions expressed are however those of the authors only and do not necessarily reflect those of the European Union. Neither the European Union nor CINEA can be held responsible for them. We would like to thank the SSH CENTRE project for giving us the opportunity to create this exciting collaboration.

References

Barich, A., Stokłosa, A. W., Hildebrand, J., Elíasson, O., Medgyes, T., Quinonez, G., Casillas, A. C., & Fernandez, I. (2022). Social license to operate in geothermal energy. *Energies, 15,* 139. https://doi.org/10.3390/en15010139

Chavot, P., Heimlich, C., Masseran, A., Serrano, Y., Zoungrana, J., & Bodin, C. (2018). Social shaping of deep geothermal projects in Alsace: Politics, stakeholder attitudes and local democracy. *Geothermal Energy, 6*(1), 1–21. https://doi.org/10.1186/s40517-018-0111-6

Chen, S., Zhang, Q., Andrews-Speed, P., & Mclellan, B. (2020). Quantitative assessment of the environmental risks of geothermal energy: A review. *Journal of Environmental Management, 276,* 111287. https://doi.org/10.1016/j.jenvman.2020.111287

Elstub, S., Carrick, J., Farrell, D. M., & Mockler, P. (2021). The scope of climate assemblies: Lessons from the Climate Assembly UK. *Sustainability, 13*(20), 11272. https://doi.org/10.3390/su132011272

Ejderyan, O., Ruef, F., & Stauffacher, M. (2020). Entanglement of top-down and bottom-up: Sociotechnical innovation pathways of geothermal energy in Switzerland. *The Journal of Environment and Development, 29*(1), 99–122. https://doi.org/10.1177/1070496519886008

European Commission. (2019). *The European Green Deal*. COM/2019/640 final. European Commission.

European Commission. (2022). *REPowerEU Plan*. COM(2022) 230 final. European Commission.

European Union. (2023). *Directive (EU) 2023/2413 of the European Parliament and of the Council of 18 October 2023 amending Directive (EU) 2018/2001, Regulation (EU) 2018/1999 and Directive 98/70/EC as regards the promotion of energy from renewable sources, and repealing Council Directive (EU) 2015/652*. Official Journal of the European Union.

Friederichs, G. (2021). *Innovative finance mechanisms for geothermal energy*. Crowdthermal. https://www.crowdthermalproject.eu/wp-content/uploads/2021/09/CROWDTHERMAL-D2.3-Innovative-Finance-Schemes-Final-v2.pdf. Accessed 25 January 2024.

Geothermal Implementation Working Group. (2023). *Implementation plan*. SETIS. https://setis.ec.europa.eu/document/download/f74cb5d6-30d8-492c-a528-9652a0d4b6c1_en?filename=Implementation%20plan%20on%20deep%20geothermal%20energy.pdf. Accessed 18 March 2024.

Gombert, P., Lahaie, F., & Cherkaoui, A. (2018). *State of knowledge about the risks, impacts and potential inconveniences associated with deep geothermal*. INERIS. https://www.ineris.fr/sites/ineris.fr/files/contribution/Documents/DRS-18-171541-05971A-RAP-risques_geothermie-v19c-english-finale%20%28003%29.pdf. Accessed 31 January 2024.

Hildebrand, J., & Klein, K. (2022). *Stakeholder and case study analysis report*. Crowdthermal. https://www.crowdthermalproject.eu/wp-content/uploads/2022/09/CROWDTHERMAL-D1.3_Stakeholder-and-case-study-analysis-report-update.pdf. Accessed 25 January 2024.

OJEU. (2023). Revised Energy Directive—Directive (EU) 2023/2413 of the European Parliament and of the Council of 18 October 2023. Document 32023L2413. European Parliament. https://eur-lex.europa.eu/legal-content/EN/TXT/?uri=CELEX:32023L2413. Accessed 31 January 2024.

Rohse, M., Barich, A., Bossennec, C., Loschetter, A., Manzella, A., Pellizzone, A., Ryder, S., & Soutar, I. (2024). Interdisciplinary workshop on societal engagement in geothermal energy. *Zenodo*. https://doi.org/10.5281/zenodo.11202258

Ryder, S. S., Dickie, J. A., & Devine-Wright, P. (2023). "Do you know what's underneath your feet?": Underground landscapes and place-based risk perceptions of proposed shale gas sites in rural British communities. *Rural Sociology, 88*(4), 1131–1162. https://doi.org/10.1111/ruso.12513

Ryder, S. S., Rohse, M., & Abesser, C. (under review). Understanding and evaluating operator engagement ethos and practices: Examples from the UK deep geothermal sector. Submitted to *Energy Policy*. Pre-peer review version available here: https://papers.ssrn.com/sol3/papers.cfm?abstract_id=4784596 Accessed 2 May 2024.

Simone, A., Pellizzone, A., Gaballo, C., & Cattaneo, M. C. (2023). Lombardy needs from citizens' voices. *Zenodo*. https://doi.org/10.5281/zenodo.7687568

The Citizens' Assembly on Climate Issues. (2021). *The Citizens' Assembly's recommendations*. Copenhagen: Danish Ministry of Climate, Energy and Utilities.

Westlake, S., John, C. H. D., & Cox, E. (2023). Perception spillover from fracking onto public perceptions of novel energy technologies. *Nature Energy, 8*, 149–158.

Open Access This chapter is licensed under the terms of the Creative Commons Attribution 4.0 International License (http://creativecommons.org/licenses/by/4.0/), which permits use, sharing, adaptation, distribution and reproduction in any medium or format, as long as you give appropriate credit to the original author(s) and the source, provide a link to the Creative Commons license and indicate if changes were made.

The images or other third party material in this chapter are included in the chapter's Creative Commons license, unless indicated otherwise in a credit line to the material. If material is not included in the chapter's Creative Commons license and your intended use is not permitted by statutory regulation or exceeds the permitted use, you will need to obtain permission directly from the copyright holder.

CHAPTER 4

Create a Co-learning Environment for Geothermal Energy Communities Across the European and African Unions

Chris Büscher, Walter Wheeler, Susan Onyango,
Jacques Varet, Fabio Iannone, Eleonora Annunziata,
Yves Geraud, and Peter Omenda

Policy Highlights To achieve the recommendation stated in the chapter title, we propose the following:

- Enable and encourage Geothermal Energy Communities (GECs) in the European Union and African Union.
- Acknowledge and embrace the potential of geothermal resources for energy communities.

C. Büscher (✉)
Department of Cultures, Politics and Society, University of Turin, Turin, Italy
e-mail: chrisbuscher@hotmail.com

W. Wheeler
Department of Energy and Technology, NORCE Norwegian Research Centre AS, Bergen, Norway
e-mail: walter.wheeler@norceresearch.no

© The Author(s) 2024
A. Crowther et al. (eds.), *Strengthening European Energy Policy*,
https://doi.org/10.1007/978-3-031-66481-6_4

- Assess GECs' feasibility using interdisciplinary and participatory approaches.
- Investigate and address critical GEC issues, including financial obstacles and land politics.
- Develop an enabling and social learning environment for GECs.
- Ensure GEC projects are embedded in the community by using on-site transdisciplinary co-learning workshops that bring together Engineering, Social Scientists, and Geoscientists as well as community representatives and critical outsiders.

Keywords Subsurface · Social development · Participatory approach · Land politics · Community engagement

S. Onyango
Laboratoire d'Anthropologie Politique, L'École des Hautes Etudes en Sciences Sociales, Paris, France
e-mail: susan.onyango@ehess.fr

J. Varet
Géo2D SARL, Orléans, France
e-mail: j.varet@geo2d.com

F. Iannone · E. Annunziata
Institute of Management, Sant'Anna School of Advanced Studies, Pisa, Italy
e-mail: fabio.iannone@santannapisa.it

E. Annunziata
e-mail: eleonora.annunziata@santannapisa.it

Y. Geraud
GeoRessources Laboratory, Université de Lorraine, Nancy, France
e-mail: yves.geraud@univ-lorraine.fr

P. Omenda
Scientific and Engineering Power Consultants (SEPCO), Nairobi, Kenya
e-mail: pomenda@sepco.co.ke

4.1 Introduction

This chapter integrates three elements important to European Union (EU) energy policy: geothermal energy (herein 'geothermal') in the energy transition; communities as a key driving agent; and EU energy relations with Africa. Whereas geothermal's share of the energy supply has long been below its potential, the European Commission (EC) now seeks to raise this share through policy incentives. This is fortunate, considering geothermal's potential to accelerate the energy transition. What is lacking, however—and what we therefore call for—is policy acknowledging geothermal's potential for local communities, similar to energy communities focused on solar and wind electricity. In short, we recommend that the EC encourage and enable geothermal-powered energy communities (GECs)—both in the EU as well as in the African Union (AU), through the EU–AU energy partnership.

Our recommendation speaks to the EC in its two roles: that of policymaker and enabler of the EU's energy transition; and as development partner helping Africa realise its 'green energy future' (AEEP, 2023). The recommendation draws on insights gained from a two-day co-learning workshop, augmented with insights from an interdisciplinary EU–AU research project called Geothermal Village. The Geothermal Village project is part of the LEAP-RE programme, and aims to introduce off-grid, geothermal-powered energy systems to four rural communities in Kenya, Rwanda, Djibouti, and Ethiopia. There is significant geothermal potential in the EU to realise the same concept—hence our call to support GECs in both continents. What is more, we believe GECs from both continents could learn from each other, a point we elaborate in the concluding section.

The co-learning workshop that informs our recommendation took place in November 2023 in Homa Hills, in western Kenya. We brought together Geoscientists, Engineers, and Social Scientists from Europe and Africa, representatives of the Homa Hills community and three Kenyan civil-society advocates with experience in EU–AU energy cooperation. The 18 participants tackled three main topics: (1) the promises and pitfalls of GECs in East Africa; (2) how these relate to potential GECs in the EU; and (3) how both could be supported in a broader framework of EU–AU energy cooperation. Each discipline/group shared its expertise and views followed by discussions. The plenary sessions were alternated with two types of subgroup sessions 'in the field', on which more below.

4.2 ENABLE AND ENCOURAGE GEOTHERMAL ENERGY COMMUNITIES

4.2.1 Geothermal's Potential for Energy Communities

Geothermal—the Earth's natural heat—is a renewable energy source with a high and diversified development potential. This potential is hardly tapped in the EU, with geothermal only providing 3% of its renewable energy (RE) (European Union, 2023). Current EU policy calls for a significant increase in geothermal by 2030, acknowledging geothermal's importance for the energy transition (Dulian, 2023; European Commission, 2023). A point that is underacknowledged is the major role that communities can play in, and multiple ways in which they can benefit from, geothermal development. We suggest geothermal can complement existing energy communities in the EU that mostly draw on solar or wind electricity.

Geothermal provides a powerful new source for energy communities, understood as "decentralised socio-technical systems where energy is jointly generated and distributed among a community of households locally" (Fouladvand et al., 2022, p. 1). As Table 4.1 shows, geothermal has some advantages over, and can therefore complement, solar and wind as RE sources. Geothermal offers continuous baseload energy at low cost, is immune to the vagaries of climate and weather, can store energy as heat or cold, and has a relatively modest footprint (Lovering et al., 2022). It offers a range of uses, depending on the geothermal source's temperature. High-heat resources (200+ °C, found at deep depths) allow for generating electricity on a large scale. On the low end, resources below 30 °C (at shallow depth) can be used to heat or cool individual buildings, typically using heat pumps. It is the medium-heat geothermal resources (50–120 °C) that offer the most potential for powering energy communities, that is, for meeting a range of community-scale energy needs at economical drilling depths. These community-scale energy needs include heating and cooling neighbourhoods, and small-scale productive uses (e.g. greenhouse farming, pasteurising and freezing dairy products). Energy wells can store surplus heat or cold and, if the temperature exceeds 90 °C, produce electricity (Onyango, 2022; Varet et al., 2014).

GECs, however, need to be promoted and require tailored incentives because most communities are unfamiliar with geothermal as a potential power source. As such, they can be wary of disturbing the subsurface

Table 4.1 Comparison of renewable energy sources, highlighting the strengths of each, and aspects unique to geothermal

RE characteristics	Geothermal		Wind	Hydro-electric	Biomass	Solar	
	<90 °C	>90 °C				Thermal	PV+ batt
Produce electricity	No	Yes	Yes	Yes	Yes	No	Yes
Produce heat	Yes	Yes	No	No	Yes	Yes	No
Long-term thermal storage	Yes	Yes	No	No	No	No	No
Up-front financial risk	In cases	Yes	No	No	No	No	No
% of year at full output	98%	98%	*50%	*98%	*98%	*50%	*98%

Asterisk '*' indicates output depends on weather/climate. 'PV+ batt' indicates solar photovoltaic electricity with battery storage

(Steward & Lewis, 2017) and expect the resources to be difficult and expensive to exploit. Fortunately, new technologies have made it easier and less costly to access these resources (Ciucci, 2023). While this has raised the prospects for GECs in the EU, these enhanced conditions have mostly led to larger private and/or public companies taking advantage of geothermal.

In large parts of the AU, especially along the East African Rift System (EARS), medium-heat geothermal resources abound and are accessible even at shallow depth, making them easier to exploit than in the EU. The problem here is that African governments and their development partners prioritise large-scale electricity production from deep, high-heat geothermal resources and are unaware of, or neglect, the potential for community energy development. Here lies a major opportunity for the EU, as the world's largest development cooperation donor, to support development of GECs in the AU.

Overall, we welcome the EC's efforts to promote geothermal, and emphasise its potential for community development in both the EU and AU. Yet, having potential is one thing, developing GECs is quite another.

4.2.2 Assessing GEC's Feasibility: An Interdisciplinary and Participatory Approach

Developing a GEC requires assessing its feasibility from different disciplinary perspectives, and in close cooperation with community members and groups. One dimension to consider is the subsurface geology and temperature, and the landscape topography in relation to community power requirements. Geoscience and Engineering studies give the community an estimate of the subsurface energy potential, the cost of drilling to access the hot water or steam, and the degree of risk to drinking water and the environment. These studies are site-specific, typically time-consuming, and present a large up-front cost. Confirmation of the resource is only obtained by drilling wells, which are themselves expensive.

A second, crucial dimension is the GECs' social aspects and dynamics. Understanding is gained by assessing a community's socio-economic needs and how a GEC can meet (some of) those. To this end, different segments of the local population should be included and represented in a GEC's decision-making process to share their visions and needs. This is a complex process, given the heterogeneity—and in some cases in the EU and AU, indigeneity—of communities. In Eastern Africa, the surface geothermal manifestations have been used by local populations since time immemorial. Their indigenous knowledge regarding the resources must, along with scientific knowledge, inform the third dimension: the engineering design and implementation of a GEC (Onyango & Varet, 2016), so that it responds to socio-economic needs and fits the community's unique context. The intersection of energy potential and social needs informs the technical design and thus informs the basis of a GEC. This requires continuous interchange with social groups locally and should be flexible to follow the evolution of the community's energy use.

Getting thorough insight into each of these dimensions and their interplay requires an approach that is both interdisciplinary and participatory. Such an approach informed the co-learning workshop we held. It not only brought into dialogue the researchers from the Geosciences, Social Sciences, and Engineering, who had each collected data on the Homa Hills prospective GEC, but also brought these scientists into dialogue with community representatives and civil-society advocates. Each discipline/group presented its unique view(s) on the promises and prospects of a GEC. This '360 degree' overview allowed a further discussion on how the different views relate. During these plenary sessions, there were two alternated subgroup sessions in the field to enrich the transdisciplinary dialogues. One subgroup session was a visit to hot springs associated with the deeper geothermal resource that would be drilled to power a Homa Hills GEC. The other comprised short transect walks in which we observed and discussed different social and economic activities in the community that could potentially be powered by geothermal energy.

Being in and jointly walking the site enabled the groups to better imagine and picture how a concrete geothermal energy system could take shape, taking into account the different aspects. For instance, we discussed a system's socio-spatial properties—that is, where wells and buildings of the geothermal energy system could best be constructed considering the soils, subsurface geothermal resource, proximity to households and businesses, land ownership, and where technical solutions (e.g. pipes, wires) would be placed. People living adjacent to the resources told us on the spot how they value the resources (hot spring waters) and use those daily. The board of a Community-Based Organisation established for the GEC shared their views on how best to approach the different development stages. The workshop thus enriched the insights and understandings of individual actors, and brought together different scientific and non-scientific perspectives that will jointly inform the next development stages of the GEC in Homa Hills. While most of this workshop was dedicated to this specific GEC, we believe the idea behind it is generally applicable to all GECs. Whether in workshop or other form, it is valuable, if not necessary, to juxtapose and integrate the different (non-)scientific perspectives.

4.2.3 Critical Issues and Barriers to GECs

The workshop, as well as our research experience on the Geothermal Village project and secondary data, also raised critical issues that need to be addressed in different development stages of GECs.

An important, practical issue relates to the up-front financial and other resources required for assessing a GEC's feasibility. For most community-sized uses, this will include expert surface exploration followed by drilling a well to confirm geothermal potential—a scenario perhaps more likely in much of the EU than in the EARS where medium-heat resources are often apparent at the surface. Drilling a well is relatively expensive and, as it may reveal insufficient geothermal potential for a GEC. There is thus a risk—although a low one compared to drilling for high-heat, deep resources—of losing the investment. GECs should be able to hedge such risk through risk-mitigation funds, which are currently only available for high-heat, deep drilling. A related challenge is the time taken for the feasibility assessment. In our Geothermal Village research, we observed that this tests the patience of community members, especially those who are less intensively involved in the preparations of a GEC.

The workshop also raised the issue of how people perceive the underground. Some community members in our research sites attribute special powers to geothermal resources, be they spiritual, religious, or medicinal (Onyango, 2022). Some also believe interfering with subsurface geothermal resources may bring harm, rather than good. Likewise, in Europe, events such as fracking, carbon sequestration, and building damage, due to subsidence or earthquakes caused by gas extraction, have made people wary of subsurface operations. Such perceptions and lived experiences need to be taken very seriously when developing a GEC—those involved should be transparent about the risks and uncertainties as well as the advantages of geothermal energy, let the decisions come from the community, and involve the community in monitoring.

Closely related are land issues. These were discussed at length in the workshop, because current land and subsurface policies and legislation in most Eastern African countries hinder effective participation of communities in geothermal development processes. Most communities lack the resources (legal, financial, technical) to compete for geothermal exploitation licences against private investors. Most governments, under the influence of liberalisation and privatisation, privilege private sector development of energy (including geothermal) resources, with severe implications for energy justice. In the Homa Hills GEC case, a private developer received the licence for geothermal exploration. The developer is favourable to the community creating a GEC, whereby the latter can use shallow geothermal resources up to a set depth, as long as the developer's licence rights are not restricted. While this enables the community to develop its GEC for now, the community still finds itself dependent on, and in an uneven power relationship with, a private developer. This is contrary to the rationale of GECs, whereby communities have autonomy over energy decision-making. Communities must, therefore, be supported to gain geothermal development licences themselves or at least be entitled to the shallow resources. Commercial licences would still be attractive to investors if the uppermost 800m (approximately) of the subsurface was reserved for GECs.

Recent research on this topic in the EU shows similar and additional factors to consider. Fouladvand et al. (2022) argue that behavioural and institutional aspects are particularly important in realising (geo)thermal energy communities. This notably includes the role of community boards in leading such energy communities. This is in line with our own research experience, where a Community-Based Organisation proved to be a key actor in the broader social arrangement of a GEC. Fouladvand et al. (2022) also call attention to the 'four A's' of availability, affordability, accessibility, and acceptability, that shape GECs' potential of becoming energy-secure and inclusive systems. While there are always trade-offs between these dimensions when developing GECs, they conclude GECs should be feasible in many places. Finally, regulations for the social and technical design of a GEC are not as clear as they are for electrical energy communities, thus requiring further development.

4.3 Achieving Our Recommendation

So far, this chapter has raised and discussed important preconditions and critical issues that need to be met or addressed for GECs to be realised. In this concluding section, we propose what EU agencies could do to enable GECs, and as such achieve the recommendation outlined in our title—*create a co-learning environment for Geothermal Energy Communities across the European and African Unions*.

Geothermal energy offers much potential for community development in the EU and AU, but this potential has been underacknowledged and untapped. We thus call on the EC—and specifically its Directorate-Generals for Energy and for International Partnerships—to address this policy gap and to *encourage and enable the development of GECs in the EU and, in its role of donor, in the AU*. GECs are a novel kind of energy community, whose feasibility and development hinge on some important preconditions and issues discussed above.

There is the need to acknowledge *and embrace the potential of geothermal resources for energy communities*. Communities should be encouraged, and enabled, to become more empowered players in geothermal development than is currently the case. This requires community-friendly policies and regulatory frameworks. These are policies and frameworks that allow for, and stimulate communities to, guide the geothermal development process, recognise their representative bodies as key stakeholders rather than peripheral players, and adjust support/aid mechanisms and the funding bureaucracy accordingly. By recognising energy communities as a crucial agent in the energy transition in the EU (European Commission, Directorate-General for Energy, 2019), and by providing support through the Energy Community Repository, the EC already took an important step. Yet, the unique character and subsurface investment of GECs require specific support and risk alleviation.

Support is especially required in the stage of *assessing a GEC's feasibility using interdisciplinary and participatory approaches*. The community's insights, knowledge, and collective needs should be central but may require EC support mechanisms for close engagement with advisory bodies to guide the social, natural, and technical scientific investigations, as well as the financial planning and implementation.

This feasibility stage should *involve the investigation of critical issues specific to GECs*. One such issue relates to finance. Whether in the EU or AU, the amount required for setting up GECs likely exceeds the financial resources available to communities. Communities who wish to develop a GEC should therefore have easier access to financial resources, with favourable conditions, such as a low interest and long payback time, and a risk mitigation fund. The European Investment Bank (EIB) could be an important facilitating agency. As one of the largest multilateral financial institutions in the world, the EIB is well-equipped to offer a financial support programme in the EU and AU—perhaps in partnership with regional investment banks and pension funds pulling out of fossil fuels. Another crucial issue for GECs is gaining land rights and licences to develop geothermal resources in the liberalised energy landscape. Current policy regimes tend to favour big, often private, players drilling for deep energy, and discourage, rather than enable, symbiotic community-scale shallower geothermal energy. An enabling environment for GECs recognises and allows alternative property-rights arrangements better suited to energy forms organised around the commons (Bridge & Gailing, 2020). The EC would do well to allow and support property regimes that are based on communal ownership and management of energy systems—whether in its own territory or (via aid) in the AU. In its role of donor, the EU could more actively encourage African governments to enable geothermal development at community level.

Finally, GECs are new and innovative undertakings, but also adventurous and uncertain. Considering this, it will help existing and prospective GECs if they were to connect and partner up in peer-learning programmes, where they could exchange insights and lessons learned. Such a programme could take different shapes. Given physical and social proximity, GECs partnering up within either the EU or the AU is most easily arranged. Yet, setting up connections between GECs from both continents can be equally rewarding and appropriate. After all, as this chapter has shown, communities in the EU and AU have similar energy community issues that merit exchange. Two platforms may take this further—one is the Africa—EU Energy Partnership (AEEP, 2023), the other is the EU-Africa Green Energy Initiative that aims to enhance what GECs offer: clean energy access via off-grid decentralised solutions (European Commission, 2022).

Acknowledgement The authors wish to thank all participants for their participation in and contribution to the workshop. This chapter was also supported by EC funding from the LEAP-RE project "Geothermal Village", Horizon 2020 Research and Innovation Programme, under Grant Agreement number 963530.

References

AEEP. (2023). *The gateway for joint action on a Green Energy future*. The Africa-EU Energy Partnership.
Bridge, G., & Gailing, L. (2020). New energy spaces: Towards a geographical political economy of energy transition. *Economy and Space, 52*(6), 1037–1050. https://doi.org/10.1177/0308518X20939570
Ciucci, M. (2023). *Innovative technologies in the development of geothermal energy in Europe*. Briefing to the European Parliament by the Directorate-General for Internal Policies.
Dulian, M. (2023). *Geothermal energy in the EU*. Briefing to the European Parliament by the European Parliamentary Research Service.
European Commission. (2022). *EU-Africa: Global Gateway Investment Package—Green Energy initiative*. European Commission. https://international-partnerships.ec.europa.eu/policies/global-gateway/initiatives-region/initiatives-sub-saharan-africa/eu-africa-global-gateway-investment-package_en
European Commission. (2023). *State of the Energy Union 2023*. COM/2023/650 final. European Commission.
European Commission, Directorate-General for Energy. (2019). *Clean energy for all Europeans*. Publications Office of the European Union.

European Union. (2023). *Directive (EU) 2023/2413 of the European Parliament and of the Council of 18 October 2023 amending Directive (EU) 2018/2001, Regulation (EU) 2018/1999 and Directive 98/70/EC as regards the promotion of energy from renewable sources, and repealing Council Directive (EU) 2015/652.* Official Journal of the European Union.

Fouladvand, J., Ghorbani, A., Mouter, N., & Herder, P. (2022). Analysing community-based initiatives for heating and cooling: A systematic and critical review. *Energy Research & Social Science, 88*, 1–18. https://doi.org/10.1016/j.erss.2022.102507

Lovering, J., Swain, M., Blomqvist, L., & Hernandez, R. R. (2022). Land-use intensity of electricity production and tomorrow's energy landscape. *PLoS ONE, 17*(7), 1–17. https://doi.org/10.1371/journal.pone.0270155

Onyango, S. (2022). Cultural uses and values placed on geothermal resources by Kenya's Luo of Homa Hills: Views from a socio-anthropological perspective. In *Proceedings, 9th African Rift Geothermal Conference, Djibouti, Djibouti city, 3–5 November 2022.*

Onyango, S., & Varet, J. (2016). Future geothermal energy development in the East African Rift Valley through local community involvement: Learning from the Maori's experience. In *Proceedings, 6th African Rift Geothermal Conference, Addis Ababa, Ethiopia, 2–4 November 2016.*

Steward, I. S., & Lewis, D. (2017). Communicating contested geoscience to the public: Moving from 'matters of fact' to 'matters of concern.' *Earth-Science Reviews, 174*, 122–133. https://doi.org/10.1016/j.earscirev.2017.09.003

Varet, J., Jecton, A., Omenda, P., & Onyango, S. (2014). The "Geothermal Village" concept: A new approach to geothermal development in rural Africa. In *Proceedings, 5th African Rift Geothermal Conference, Arusha, Tanzania, 29th–31st October 2014.*

Open Access This chapter is licensed under the terms of the Creative Commons Attribution 4.0 International License (http://creativecommons.org/licenses/by/4.0/), which permits use, sharing, adaptation, distribution and reproduction in any medium or format, as long as you give appropriate credit to the original author(s) and the source, provide a link to the Creative Commons license and indicate if changes were made.

The images or other third party material in this chapter are included in the chapter's Creative Commons license, unless indicated otherwise in a credit line to the material. If material is not included in the chapter's Creative Commons license and your intended use is not permitted by statutory regulation or exceeds the permitted use, you will need to obtain permission directly from the copyright holder.

PART III

Navigating Knowledges for the Built Environment

CHAPTER 5

Facilitate the Development of Energy Literacy Amongst Citizens to Support Their Meaningful Participation in the Energy Transition

Philippa Calver, *Ami Crowther*, *and Claire Brown*

Policy Highlights To achieve the recommendation stated in the chapter title, we propose the following:

- Ensure that citizens have appropriate knowledge to meaningfully participate in the energy transition by providing accessible information that reflects citizens' contexts.

P. Calver (✉) · A. Crowther
Global Sustainability Institute, Anglia Ruskin University, Cambridge, UK
e-mail: philippa.calver@aru.ac.uk

A. Crowther
e-mail: ami.crowther@aru.ac.uk

C. Brown
The Tyndall Centre for Climate Change Research, The University of Manchester, Manchester, UK
e-mail: claire.brown-3@manchester.ac.uk

© The Author(s) 2024
A. Crowther et al. (eds.), *Strengthening European Energy Policy*,
https://doi.org/10.1007/978-3-031-66481-6_5

- Consider the framing of information to support participation in the energy transition, including the broader impact, and relationships of energy transitions with other aspects of everyday life.
- Draw upon existing networks, independent intermediaries, and communication channels to build trust in the information provided.
- Bring together Social Sciences and Humanities (SSH) and more technical researchers to explore potential energy futures and the diverse knowledge required for citizens to meaningfully participate in, and benefit from, these energy futures.

Keywords Energy literacy · Knowledge · Framing · Trust · Participation

5.1 Introduction

The European Union (EU) has ambitions to achieve carbon neutrality by 2050. The realisation of this ambition is supported by a portfolio of energy policies within REPowerEU that cover the decarbonisation of the energy system, the reduction of energy demand particularly at peak times, energy efficiency, and tackling energy security (European Commission, 2022). The enactment of these policies will see significant, and rapid modifications to the energy system at all scales. Residential changes will include increases in energy storage, residential energy production, demand-side flexibility schemes, energy-related technologies, and community energy schemes. We argue that successful energy system modifications require societal engagement and collaboration between citizens, objects, and actors (such as governments or private businesses) via discrete projects, the energy market, and everyday social practices. Drawing upon the complexity of ways in which citizens can engage with energy systems (Chilvers et al., 2018), we collectively term these forms of engagement: 'participation'.

This chapter focuses on the development of citizens' energy system participation, and the need for energy-related knowledge, or energy literacy. Whilst multiple confounding factors influence citizens' energy system participation, our prioritisation of knowledge is due to evidence that participation and the gaining of benefits from energy system changes can be inhibited when knowledge is missing. Furthermore, a lack of, or incorrect, information[1] has been shown to lead to non-optimal outcomes for the electrical network when citizens engage with new technologies and market offers (Calver et al., 2022). EU citizens need to be energy literate, which we argue must go beyond knowing how to save energy within the home. They must understand how to meaningfully participate in this transition—for example, through the adoption or rejection of new technologies or tariffs—via involvement with community schemes, and through energy system trials and co-creation. Importantly, they must have information on how likely they are to benefit from these different forms of participation based on their circumstances, allowing them to better navigate the ever-increasing complexity of the energy market, and ensuring informed consent for participation. These are themes discussed within this chapter, with a focus on the types and details of knowledge that citizens require, in addition to the conditions required for this knowledge to be developed.

As an interdisciplinary group (Human Geography, Engineering, and Construction Project Management), we brought together knowledge of the proposed technical changes to centralised energy system within the EU, with Social Sciences and Humanities (SSH) understandings of how citizens are expected to, and do, participate in energy systems—marrying together two often discrete areas. Our policy recommendations were informed by a workshop in Greater Manchester, UK, that brought together different actors associated with low-carbon energy transitions (including from the public sector, academia, and the charity sector) and those from different disciplinary backgrounds (including Economics and Environmental Science). During the workshop, the interdisciplinary research team adopted the dual roles of researcher and workshop participant. The workshop highlighted where different disciplinary and professional perspectives converged and provided examples and initiatives to reflect upon.

The workshop discussions focused on: (1) technical energy system interventions; and (2) the role of citizens in these interventions. More technical-leaning discussions were encouraged through sessions that looked back at energy system changes over time, with the subsequent discussion looking forward into the energy systems of the future, drawing upon energy scenarios as prompts. More social-leaning reflections were

---

[1] Information pertains to factual knowledge, whilst advice consists of recommendations of potential actions.

prompted when discussing the opportunities and challenges of these changes, with a focus placed on the consequences for different individuals. During the discussions, facilitators helped ensure that knowledge and the conditions needed for the development of knowledge were considered.

5.1.1 The Knowledge and Conditions for Developing Energy Literacy

There is a plethora of research on energy literacy, and the development of knowledge, attitudes, and behaviours to allow citizens to engage in energy decision-making (Santillán & Cedano, 2023; van den Broek, 2019). Predominantly, this work holds the narrative that imparting information and developing energy-literate citizens will lead to more rational energy decisions, principally leading to reduced energy consumption and the taking up of specific low-carbon technologies. However, findings linking energy literacy to energy behaviour change are mixed (Adams et al., 2022), and we advance that the premise of a 'rational' or 'right' action no longer stands where everyday citizens are expected to play an active role in an energy system beyond using or curtailing energy. Instead, with the ever-increasing complexity of the energy market, we contend that a greater focus should be placed on ensuring citizens have the diverse knowledge needed to confidently navigate and meaningfully participate in the new energy landscape.

Whilst there is no specific EU policy related to energy literacy, there is an awareness of the importance of developing energy literacy amongst the EU and Member States. For example, the Hungarian National Energy and Climate Plan includes a commitment to establish energy/climate literacy-enhancing campaigns and educational measures (Hungarian Ministry of Innovation & Technology, 2021). Similarly, the revised EU Energy Performance of Buildings Directive considers the provision of information, such as Energy Performance Certificates, as a measure to improve energy literacy (European Union2024). Several EU-funded projects have also engaged with the concept. These projects have developed practical tools aiming to increase citizen awareness and knowledge of how they can participate in the energy system, with many projects focusing on energy communities. For example, the Horizon2020 NEWCOMERS project (Drevenšek & Tajnšek, 2022) advocated for education to support change, with citizens' understanding a prerequisite for meaningful participation in energy communities. Similarly, the REScoop network has co-created a handbook for supporting citizens and local communities to participate in community-led renewable change projects (Friends of the Earth Europe, 2020).

The academic literature suggests energy literacy consists of different knowledge areas. van den Broek (2019) extrapolates four aspects of energy literacy knowledge: device energy literacy (understanding how specific technological artefacts work); action energy literacy (understanding options available in relation to changing energy practices to reduce energy consumption); financial energy literacy (understanding short- and long-term costs and potential gains from different forms of energy system participation); and multifaceted energy literacy (understanding energy systems more broadly, and how their action is part of a broader system). When discussing the knowledge required for citizens to make informed decisions about participating in the energy system, during the workshop, all four aspects of energy literacy were implicitly referred to by participants. There was a recognition of the varied ability of citizens to participate in, make decisions about, and represent themselves within the energy system depending on their context.

Informed by the workshop discussions, existing literature, and the different disciplinary perspectives of the research team, we have identified three areas that can be addressed through policy to facilitate citizens to develop their energy literacy—(1) tackling the information gap between generic and bespoke provision, (2) framing participation in the energy system beyond energy, and (3) ensuring citizen access to trusted and trustworthy actors and sources of information. These three areas inform and dictate how individuals receive, process, and understand information related to energy decision-making.

5.1.2 *Tackling the Information Gap Between Generic and Bespoke Provision*

To ensure that citizens have appropriate energy-related knowledge to meaningfully participate in the energy transition, there is a need to ensure that the information provided is reflective of the different contexts of citizens. Much of the available information about opportunities and outcomes of citizen participation in the energy transition (e.g. the adoption of new technologies or engagement with demand-side response schemes) is generic in nature. For example, providing the average payback times for LED lamps; TV campaigns about new energy tariffs; or generic guidance on how to most effectively use heat pumps based on average properties and routines. The drawbacks of more generic information provision are shown in research, with scholars highlighting how this can lead to households making decisions on incomplete, and sometimes incorrect, information based on their circumstances (Fell et al., 2014). As stated by Krishnamurti et al. (2012, p. 796), there is the need to communicate "realistic expectations of benefits and risks, explicitly addressing the misconceptions commonly found in the mental models of consumers forced to rely on the information currently available to them". Whilst there are often mechanisms to support more bespoke information being shared with citizens (i.e. home energy audits, renovation surveys, and plans), these may still rely on this generic guidance and are resource intensive.

The provision of tailored information on residential energy technologies can be facilitated by local initiatives, coordinated by actors such as local government or third-sector organisations. Yet, this information needs to capture the plurality of options available, so that citizens have a complete understanding of the technologies or participation options that may be suitable for them, rather than only being informed about options that are of interest to third parties (Calver et al., 2022). These local initiatives require support and resources from both EU and Member State Governments, so that they can be undertaken and to ensure they align with broader priorities. Such support and resources extend beyond finance, to include technological guidance and information for those actors coordinating local initiatives. Insights from residential energy intervention trials in the public and private sectors do not always flow through to those on-the-ground. There is a need to ensure that state-of-the-art knowledge on the opportunities and suitability of different energy system engagements can be utilised by those supporting citizens to understand their options (thus contributing to citizen knowledge), such as energy auditors, technology installers, energy advisors, social housing providers, and renovation professionals. In doing so, this will add to a growing pool of knowledge on how physical characteristics of homes, household makeup, routines, and energy needs interplay with outcomes (both for the household and the energy system), and how these relate to meeting EU policy aims. The provision of this more tailored information will help citizens gain the knowledge to better participate in the low-carbon energy transition.

5.1.3 Framing Participation in the Energy Transition Beyond Energy

The way information is framed, can influence how citizens interpret, comprehend, and respond to energy-related data and advice. A workshop participant who supports vulnerable households in implementing energy efficiency measures reflected, "people like to understand the reason behind why you need to do something". This aligns with Mert's (2008) reflections on the link between consumers understanding the broader impacts of using smart technologies and being motivated to adopt them. The justification of the energy transition and technologies also provides the opportunity to make links across a range of priorities for individuals. For example, a participant from Greater Manchester's regional government gave the example of the recently launched Local Energy Advice Demonstrator (LEAD) in Greater Manchester, which focuses on the financial benefits of home energy efficiency improvements (Groundwork, 2024). As such, the justification of policy priorities can support knowledge generation and participation.

Acknowledging the interplay between participating in the energy system and other aspects of daily life is pertinent not only for citizens, but also for the actors involved in providing energy advice. This is particularly evident when reflecting upon the potential health consequences of household participation in the energy system. Those providing advice or supporting activities need to be aware of the consequences of poor decision-making and the potential negative impacts on households. For example, workshop participants gave powerful accounts of health impacts as a co-benefit or conflict, alongside energy and financial outcomes. Accounts included discussion of where houses were adapted to be low-carbon, but resulted in less liveable spaces (e.g. overheating, an increase in condensation), which affected those individuals' acceptance of these new energy practices, and had a knock-on effect on influencing acceptance of those within their networks. Thus, there is a need for those providing advice and support to understand these health consequences, and how best to communicate these, alongside enhanced consumer protection legislation to support the achievement of liveable homes. As such, there is also a need to think about framing and its consequence on the achievement of policy ambitions at a range of scales including that of the EU, Member States, and municipalities.

5.1.4 Ensuring Citizen Access to Trusted and Trustworthy Actors and Sources of Information

For citizens to develop energy literacy, there is a need for access to trusted sources of information. Trust was a key theme within the workshop discussions, used almost synonymously with the transparency of information and advice. This reinforces research showing the importance of energy information and advice being presented by trusted actors (Khuc et al., 2023), perceived to have positive intentions (Greenberg, 2014). Participants reflected on who is considered a trusted voice, stating both neighbours and established, reputable organisations.

Participants from the municipal government and third-sector representatives shared experiences of renovation policies and initiatives in Greater Manchester. They reflected on the importance of being up-front when sharing information and advice, particularly around potential household disruption, managing households' expectations of the process, and providing clarity on why citizens are eligible for different funding sources. Participants shared anecdotes of individuals turning down legitimate support due to a lack of trust in the organisations involved. Suggested reasons for not trusting organisations included concerns of vested interests influencing information provision and previous negative experiences with financial support schemes.

The discussions reinforced arguments for using local, trusted organisations, based within the community to articulate information about energy policies, changes in the energy system, and opportunities for citizen participation (Chambers et al., 2022). The importance of the EU, Member States, and municipalities as coordinators and enablers was highlighted during the workshop, as these different institutions can create the right policy frameworks and funding models to ensure citizens get this locally specific information from these trusted sources.

Overall, whilst the European Commission recognises the importance of building trust to support citizen participation (European Union, 2023), our workshop discussions and the literature show the importance of using multiple channels to share information on energy policy and opportunities for citizens to participate in the energy system. These multiple channels include trusted actors, particularly those with political, technological, and product impartiality.

5.2 Achieving Our Recommendation

With current strategies for achieving the EU's climate neutrality placing value in having knowledgeable and engaged citizens, the approach to energy literacy presented in this chapter can guide policy mechanisms towards this aim. This chapter calls for a refocus on developing energy-literate citizens away from simply understanding how to reduce energy in the home, towards allowing citizens to meaningfully participate in the low-carbon energy transition, and gain benefits from doing so. The achievement of our policy recommendation, as outlined in our title—*facilitate energy literacy amongst citizens to support their meaningful participation in the energy transition*—can be supported by the following sub-recommendations related to the information shared, how it is communicated, and who is sharing it.

Firstly, we discussed the use of 'average' citizen outcomes from different forms of participation within the energy system, and the need for the information provided to better reflect the different contexts of citizens. There is a plethora of data coming from trials and projects across the EU, but there is a need for intervention at national and subnational levels to collect, interpret, and communicate this information. An understanding of local contexts would allow citizens (and those actors involved with this transition) greater knowledge of potential implications, positive or negative, of energy system participation.

Secondly, we discussed how being energy literate requires citizens to have knowledge that extends beyond energy technologies and practices, such as knowledge related to finance and health. Financial implications are often considered, but knowledge on health implications needs to be further considered, along with implications on other aspects of everyday life. Whilst knowledge within these areas may increase participation, focusing on these areas also facilitates increased informed consent for those embarking on different forms of participation. The acknowledgement of the breadth of knowledge required to be energy literate is critical for policies that set out a role for individuals, such as the Renovation Wave (European Commission, 2020) and for diversifying the energy mix. Furthermore, there is value in grounding these low-carbon ambitions within defined contexts to support motivation, demonstrating the value of establishing sub-EU scale narratives within policy ambitions.

Lastly, we discuss how this information must come from trusted and trustworthy actors. We highlight how trust can be eroded by a lack of access to unbiased sources of information, a lack of clarity on what support may be available, and perceived or real vested interests. There is a need to ensure that trusted voices are involved in the articulation of information and opportunities related to shifting residential energy systems (e.g. local initiatives related to the Social Climate Fund), with the EU, Member States, and municipalities having a role in amplifying these trusted voices.

The growing role of citizens in achieving the EU's climate ambitions demonstrates the need to ensure that European societies are energy literate and have the knowledge to meaningfully participate in energy system change. The EU can take a proactive role in developing policy to facilitate the development of energy literacy and support Member States to take domestic action to support the achievement of a climate-neutral Europe. This will, however, require continued collaboration between science and technology researchers and practitioners, who are involved with envisioning and developing our future energy system, with SSH researchers and practitioners, who understand the lived experience of energy on-the-ground.

Acknowledgements We'd like to thank the participants at our workshop in Manchester for sharing their views and perspectives with us. Thank you also to the reviewers for supporting this chapter.

References

Adams, J., Kenner, A., Leone, B., Rosenthal, A., Sarao, M., & Boi-Doku, T. (2022). What is energy literacy? Responding to vulnerability in Philadelphia's energy ecologies. *Energy Research & Social Science, 91*, 102718. https://doi.org/10.1016/j.erss.2022.102718

Calver, P., Mander, S., & Abi Ghanem, D. (2022). Low carbon system innovation through an energy justice lens: Exploring domestic heat pump adoption with direct load control in the United Kingdom. *Energy Research & Social Science, 83*, 102299. https://doi.org/10.1016/j.erss.2021.102299

Chambers, J., Robinson, C., & Scott, M. (2022). *Digital inclusion in the energy system: How do we ensure the opportunities and benefits of digitalisation can be accessed by everyone?* University of Bristol. https://www.bristol.ac.uk/media-library/sites/policybristol/briefings-and-reportspdfs/2022/PolicyBristol_PolicyBriefing121_Digital_Inclusion_Energy_Robinson.pdf

Chilvers, J., Pallett, H., & Hargreaves, T. (2018). Ecologies of participation in socio-technical change: The case of energy system transitions. *Energy Research & Social Science, 42*, 199–210. https://doi.org/10.1016/j.erss.2018.03.020

Drevenšek, M., & Tajnšek, I. (2022). *Energy literacy for energy communities.* Newcomers—H2020. https://pressbooks.pub/handbooknewcomers/#:~:text=Welcome%20to%20Energy%20Literacy%20for,role%20in%20the%20energy%20system.

European Commission. (2020). *A renovation wave for Europe—Greening our buildings, creating jobs, improving lives.* COM(2020) 662 final. European Commission.

European Commission. (2022). *REPowerEU.* COM(2022) 230 final. European Commission.

European Union. (2023). *Regulation (EU) 2023/955 of the European Parliament and of the Council of 10 May 2023 establishing a Social Climate Fund and amending Regulation (EU) 2021/1060.* Official Journal of the European Union.

European Union. (2024). *Directive (EU) 2024/1275 of the European Parliament and of the Council of 24 April 2024 on the energy performance of buildings (recast).* Official Journal of the European Union.

Fell, M. J., Shipworth, D., Huebner, G. M., & Elwell, C. A. (2014). Exploring perceived control in domestic electricity demand-side response. *Technology Analysis & Strategic Management, 26*(10), 1118–1130. https://doi.org/10.1080/09537325.2014.974530

Friends of the Earth Europe. (2020). *Community energy, a practical guide to reclaiming power.* Friends of the Earth Europe, REScoop, EnergyCities. https://www.rescoop.eu/uploads/Community-Energy-Guide.pdf

Greenberg, M. R. (2014). Energy policy and research: The underappreciation of trust. *Energy Research & Social Science, 1*, 152–160. https://doi.org/10.1016/j.erss.2014.02.004

Groundwork. (2024). *Local energy advice demonstrator programme.* Groundwork. Retrieved 18 April 2024, from https://www.groundwork.org.uk/greatermanchester/gm-about/gm-our-programmes/local-energy-advice-demonstrator-programme/

Hungarian Ministry of Innovation and Technology. (2021). *National energy and climate plan.* European Commission. https://energy.ec.europa.eu/system/files/2022-08/hu_final_necp_main_en.pdf

Khuc, Q. V., Tran, M., Nguyen, T., Thinh, N. A., Dang, T., Tuyen, D. T., Pham, P., & Dat, L. Q. (2023). Improving energy literacy to facilitate energy transition and nurture environmental culture in Vietnam. *Urban Science, 7*(1), Article 1. https://doi.org/10.3390/urbansci7010013

Krishnamurti, T., Schwartz, D., Davis, A., Fischhoff, B., de Bruin, W. B., Lave, L., & Wang, J. (2012). Preparing for smart grid technologies: A behavioral decision research approach to understanding consumer expectations about smart meters. *Energy Policy, 41*, 790–797. https://doi.org/10.1016/j.enpol.2011.11.047

Mert, W. (2008). *Consumer acceptance of smart appliances: A report prepared as part of the EIE project Smart Domestic Appliances in Sustainable Energy Systems (Smart-A)*. Intelligent Energy Europe. https://ifz.at/sites/default/files/2021-02/D5_5-Consumer%20acceptance.pdf

Santillán, O. S., & Cedano, K. G. (2023). Energy literacy: A systematic review of the scientific literature. *Energies, 16*(21), Article 21. https://doi.org/10.3390/en16217235

van den Broek, K. L. (2019). Household energy literacy: A critical review and a conceptual typology. *Energy Research & Social Science, 57*, 101256. https://doi.org/10.1016/j.erss.2019.101256

Open Access This chapter is licensed under the terms of the Creative Commons Attribution 4.0 International License (http://creativecommons.org/licenses/by/4.0/), which permits use, sharing, adaptation, distribution and reproduction in any medium or format, as long as you give appropriate credit to the original author(s) and the source, provide a link to the Creative Commons license and indicate if changes were made.

The images or other third party material in this chapter are included in the chapter's Creative Commons license, unless indicated otherwise in a credit line to the material. If material is not included in the chapter's Creative Commons license and your intended use is not permitted by statutory regulation or exceeds the permitted use, you will need to obtain permission directly from the copyright holder.

CHAPTER 6

Support Place-Based and Inclusive Supply Chain, Employment and Skills Strategies for Housing-Energy Retrofit

Rachel Macrorie, Hadi Arbabi, Will Eadson, Richard Hanna, Kaylen Camacho McCluskey, Kate Simpson, and Faye Wade

Policy Highlights To achieve the recommendation stated in the chapter title, we propose the following:

- Member States should empower municipalities with resources and training to develop Building Renovation Plans supported by One-Stop Shops focused on inclusive local supply chain development, employment and skills priorities, as well as serving housing retrofit consumers.

R. Macrorie · W. Eadson
Centre for Regional Economic and Social Research, Sheffield Hallam University, Sheffield, UK
e-mail: r.macrorie@shu.ac.uk

W. Eadson
e-mail: w.eadson@shu.ac.uk

© The Author(s) 2024
A. Crowther et al. (eds.), *Strengthening European Energy Policy*,
https://doi.org/10.1007/978-3-031-66481-6_6

- Municipalities should use procurement frameworks and Direct Labour Organisations to ensure a pipeline of retrofit work and support training for good quality employment.
- Member States should implement licensing or minimum competency standards for housing retrofit professionals, ensuring certification schemes encompass a wider range of skills.
- Retrofit is an opportunity to enable new groups to enter the construction sector. Municipalities should co-create partnerships alongside employees, and support unionisation, to promote training and work opportunities for women and minorities.
- Developing inclusive pathways to a skilled housing-energy retrofit workforce is a socio-technical problem, requiring insights from social, policy, building and engineering disciplines, because retrofit interweaves human and technical practices and processes.

Keywords Renovation governance · Local-area planning · Construction skills · Capabilities · Energy efficiency

---

H. Arbabi
School of Mechanical, Aerospace, and Civil Engineering, University of Sheffield, Sheffield, UK
e-mail: h.arbabi@sheffield.ac.uk

R. Hanna
Centre for Environmental Policy, Imperial College London, London, UK
e-mail: r.hanna@imperial.ac.uk

K. C. McCluskey
Energy Futures Lab, Imperial College London, London, UK
e-mail: kaylen.camacho-mccluskey22@imperial.ac.uk

K. Simpson (✉)
Centre for Sustainable Construction and Retrofit, Nottingham Trent University, Nottingham, UK
e-mail: kate.simpson@ntu.ac.uk

F. Wade
School of Social and Political Science, University of Edinburgh, Edinburgh, UK
e-mail: Faye.Wade@ed.ac.uk

6.1 Introduction

The European Union's (EU) building stock is responsible for 36% of its greenhouse gas emissions (EEA, 2023), partly due to poor energy efficiency: a third was built before the introduction of thermal insulation regulations in the 1970s (European Commission, 2020). Most of these buildings will still be in use in 2050. Energy retrofitting residential buildings, which includes improving building fabric and moving to zero carbon heating, cooling, ventilation and electricity services, is therefore essential. However, across the EU, retrofits achieving at least 60% energy reduction are performed in only 1.2% of the building stock per year (European Commission, 2020). To address this, the revised Energy Performance of Buildings Directive (EPBD) requires Member States to create national targets for reducing energy use in domestic buildings and establish Building Renovation Plans to attain a zero-emission building stock by 2050 (European Union, 2024). However, a 'retrofit revolution' will not be achieved unless significant supply-side challenges, including labour and skills shortages, are addressed (European Commission, 2021).

EU supply chains for building energy retrofits are fragmented, typified by micro-enterprises, and insufficient workers with the requisite skills and competences to perform high-quality renovations (Renovate Europe & E3G, 2022). Instead, low skills, low demand for training, and low quality predominate in building trades where profit margins are small (Killip, 2020). The construction sector has an inadequate supply of vocational education and training (VET) and, where available, such training can lack quality and accessibility (European Construction Sector Observatory, 2020). However, it is unhelpful to focus on skills training without considering policies that shape how the VET system functions (Stroud et al., 2024).

In addition, there is a need to tackle acute diversity and inclusion challenges within the sector. Women, minority ethnic groups, and people with disabilities are hugely under-represented in the construction and retrofitting workforce, which is ageing and struggles to attract younger generations (CEDEFOP, 2023). For example, just 10% of the EU construction sector is classed as female (European Commission, Directorate-General for Internal Market, Industry, Entrepreneurship and SMEs, 2023), while in the United Kingdom (UK), only 6% are categorised as Black or minority ethnic groups and 6% are people with disabilities (CITB, 2024). Consequently, construction cultures remain male-dominated and characterised by masculine values, narratives and norms (Clegg et al., 2023). This, coupled with poor job security, challenging working conditions and health and safety concerns, can make construction careers unappealing to historically under-represented groups (European Construction Sector Observatory, 2020). A greatly expanded, more diverse and competent labour force is required to meet EPBD objectives (Renovate Europe & E3G, 2022).

This chapter therefore focuses on developing inclusive pathways to a skilled retrofit workforce. This is fundamentally a socio-technical problem: understanding the nature of the challenge and producing recommendations for change requires insights from social, policy, building, and engineering disciplines, because retrofit interweaves human and technical inputs and processes. Further, retrofitting at scale requires knowledge of interlinked social and technical systems across varied systems of provision, distinct geographies, and built environments. As such the chapter draws together academic and industry literature with insights from a workshop with academics and practitioners from the UK and EU. Participants included representatives from municipalities, not-for-profit cooperatives, charities, standards bodies, and tradespeople. The workshop structure was developed using interdisciplinary perspectives spanning Civil and Structural Engineering, Construction Management, Human Geography and Sociology. It was designed to explore multiple aspects of supply chain development including: funding and delivery models, multi-actor partnerships, and geographical scales for intervention. Workshop materials including summary notes and participant expertise are available open access (Macrorie et al., 2024).

The chapter is structured according to two themes emerging from the workshop: (1) supply-side coordination through place-based organisations, networks and One-Stop Shops (OSS) ; and (2) ensuring quality through workforce regulation and training. Both themes focus on how different approaches can be used to support inclusivity in skills provision and employment outcomes. Using these themes, the chapter develops policy recommendations for (re)training programmes and initiatives to enable the growth of a diverse, skilled building retrofit workforce.

6.2 Initiatives to Support Workforce Development

6.2.1 Supply-Side Coordination Through Place-Based Action

Place-based approaches and local partnerships are needed to mobilise and organise the supply chain for energy retrofit (Brocklehurst et al., 2022).

Municipalities and social housing landlords have housing portfolios that are geographically co-located and can act as a test bed for growing supply chain capacity (Cauvain et al., 2018). Municipalities, with their knowledge of local building stock and visibility, are well-placed to create local Building Renovation Plans but this requires resource and training to develop capabilities in, for example, analysing building stock data (Wade et al., 2022). The capacity building programme *Renocally*, for example, supports Bulgarian, Romanian, and Slovakian municipalities through technical assistance and increasing overall knowledge (BPIE, 2024). Building Renovation Plans can provide a reliable pipeline of work, which is crucial for local tradespeople and SMEs to justify investment in training. They can also be used to align local training initiatives with expected retrofitting tasks, to ensure a steady flow of appropriate on-the-job training (Topriska et al., 2018).

Where they own building stock, municipalities can procure contractors to deliver large volumes of retrofitting work and require quality guarantees. Procurement frameworks can be used to support apprenticeships, raise quality standards through setting training requirements, and support involvement from a range of contractors (Green, 2016). The public sector can also develop supply chains through Direct Labour Organisations (DLOs), like *City Building* in Scotland which directly employs 2,200 workers. DLOs can support good working conditions and inclusivity, but also create a well-trained workforce with sufficient capacity to perform retrofitting at scale.

Local intermediaries can connect formal retrofit skills and informal knowledge. For example, builders' and plumbers' merchants can support knowledge sharing among tradespeople (Wade et al., 2016). Informal and voluntary networks, like *Civic Square* in the UK, can support 'beyond-market' pathways (Galvin & Sunikka-Blank, 2014) through skills development for community-led retrofit or reciprocal repair and maintenance. These alternative networks should be supported, for instance through finance and facilitation linked to OSS initiatives. The recast

EPBD includes provision for OSSs across Europe: these initiatives can act as crucial intermediaries between supply and demand, supporting skills development (see Table 6.1). However, there is a risk that OSSs will tend towards becoming consumer-focused hubs and not realise their full potential in tackling supply-side challenges: OSS design should be equally focused on inclusive supply chain, employment, and skills priorities.

To address diversity, more tailored trade networks like *Her Own Space* (UK) and Tradeswomen Building Bridges (North America; see CIOB, 2022) can provide access to construction careers, business opportunities, and knowledge sharing for under-represented groups. More formally, trade unions can work to support diversity, protect employees, and ensure job quality and health and safety (Clarke et al., 2017). For example, the *Union of Construction, Allied Trades Technicians* (UCATT) set up a *Women's Network Forum* and a *Women in Construction Newsletter* in 2014 (Clarke et al., 2017). Trade unions could also be actively included in retrofit planning (e.g. through representation on OSS boards), and opportunities for dedicated worker networks and unionisation should be encouraged as part of place-based action.

Table 6.1 Examples of supply chain support in One-Stop Shops

One-stop shop name	Country	Business structure	Engagement with supply chains	Further information
BetterHome OSS	Denmark	Industry-led	Delivers training to existing contractors and ensures quality of service	Marmolejo-Duarte et al. (2022)
Pass Renovation	France	Regional authority	Provides access to training and facilitates business networks	www.pass-renovation.hautsdefrance.fr
OSS Oktave	France	Regional authority	Requires clients to hire certified retrofit tradespeople	Oktave (2024)
Retrofit Skills Centre	United Kingdom	Partnership of municipalities	OSS focused on retrofit skills and training	https://retrofitskills.org
RetrofitWorks	United Kingdom	SME cooperative	Sharing knowledge and skills among contractors	https://retrofitworks.co.uk

6.2.2 Ensuring Quality in the Supply Chain: Regulation and Training

Overall competence for effective energy retrofit requires a combination of manual skills, theoretical and applied knowledge, and ethical conduct (Killip, 2020). This includes competency to consider the whole building and mitigate any potential cause of poor energy performance (CORDIS, 2023), and in-depth theoretical, technical, and interdisciplinary knowledge is often needed (Clarke et al., 2020).

Sitting alongside EPBD, the EU Energy Efficiency Directive requires Member States to develop certification schemes and/or equivalent qualification for workers providing energy efficiency audits, improvements or services (European Union, 2023). These schemes could be strengthened by detailing trading licence requirements and minimum qualification standards, updated frequently (Killip, 2020). There are already examples of licensing to trade and minimum training requirements, including certification schemes that are integrated within OSSs. For example, in Czechia, energy auditors are legally mandated to complete Ministry of Industry and Trade training courses on an ongoing basis to retain their licence to practise, while in Austria a professional network of energy consultants is self-regulated, in close cooperation with regional further education authorities and a national working group (Renovate Europe & E3G, 2022). Active support of accreditation, which can build consumer confidence, should grow as technologies like heat pumps and batteries increase the need for more complex whole-system approaches and installations (Regnier et al., 2023).

However, inconsistent training content and curricula for retrofit across Europe means that skills and competencies vary. The *EU Skills Registry* (Geonardo, 2024) allows comparison of skills and competency profiles for different jobs in construction and building energy efficiency and attempts to show international equivalence of qualification and training schemes. While standards-based curricula for VET that acknowledge a more holistic perspective are developed in some regions, e.g. in Belgium and Ireland, these are rare and the majority of training still focuses on imparting siloed skillsets (Clarke et al., 2020).

A further challenge exists with frequent mismatches between the duration of apprenticeships, or on-the-job training modules, and project timelines. There is need for innovative solutions such as employers sharing responsibility for apprenticeships and vocational training through shared apprenticeship schemes (Bieler et al., 2019). Creating learning opportunities that suit worker schedules and practices is essential, especially since 93% of EU construction organisations are micro-enterprises (fewer than 10 employees) which have little flexibility to take time out of work (European Commission, 2023). Examples of best practice include *The Green Register* and *People Powered Retrofit* (both UK), which include training costs when bidding for retrofit works, while *Your Energy Your Way* (UK) ensures training is paid for in line with the UK Living Wage.

Furthermore, acceptable working conditions, including fair wages and job security, need to be enabled. For example, long hours, inaccessible sites and recruitment practices based on word-of-mouth rather than qualifications are not inclusive (Clarke et al., 2015). Outreach campaigns promoting the attractiveness of the industry and seeking to overcome cultural barriers to participation, as in the *European Construction Blueprint* and *Women Can Build* project (which runs across Spain, Germany, Portugal, Belgium, France and Italy), can provide a start in overcoming these challenges. Incorporating retrofit-related content into school curricula can also provide early exposure to the sector, influencing career decisions as early as primary school (Crespo Sánchez et al., 2023). Resources developed by the *Construction Blueprint* could be adopted for use in schools.

6.3 Achieving Our Recommendation

As per the title of this chapter, our core recommendation is that policy should *support place-based and inclusive supply chain, employment and skills strategies for housing-energy retrofit*.

The EU has adopted the revised EPBD, and Member States are required to deliver on its provisions. This chapter has emphasised the urgency and need to create a diverse, appropriately qualified retrofitting workforce to meet EPBD aims. The chapter opened with specific actions (see Policy highlights) to help achieve these based on academic and industry literature with insights from a workshop with academics and practitioners from the UK and EU.

Place-based networks and coalitions of organisations are important routes to skills development and can support diversity and inclusion in supply chains. Public sector-led approaches like DLOs, cooperatives and community-led initiatives should be supported to encourage broader engagement in retrofit skills in and beyond formal employment and enterprise. OSSs can be a useful tool for meeting some of these goals but must have a shared focus on supply as well as demand, with principles for supply chain development, employment and skills embedded within their design.

Member States should develop an ongoing licence to trade, based on minimum competency standards and more standardised and comparable qualifications. This should be a requirement of certification schemes, encompassing a holistic concept and delivery of retrofit training, raising mutual understanding between separate trades. Retrofit includes multiple technology integration into existing buildings, plus the related social processes. Therefore, an understanding of both is required to deliver good training and enabling policy. The collaboration informing this chapter was enabled through social and technical researchers working together and was supported by the expertise of diverse workshop attendees.

Working and training opportunities in building retrofit have not supported diversity due to working conditions, insufficient promotion and formal recruitment opportunities and masculine cultures. Yet, diversity could enable quicker progress on creating a more sustainable built environment. Partnerships, unions and trade networks all offer potential to support women and minorities of all ages in construction by promoting diversity and supporting better working conditions. Sufficiently professionalising the workforce will involve creating pathways for ongoing learning and career advancement that fit with varied working practices and timetables. Offering clear, place-based trajectories, including influencing career decisions in schools, can contribute to developing a skilled and more diverse workforce.

Developing inclusive pathways to a skilled housing-energy retrofit workforce is fundamentally a socio-technical problem: understanding the nature of the challenge and producing recommendations for change requires insights from social, policy, building and engineering disciplines, because retrofit interweaves human and technical inputs, practices and processes. Further, retrofitting at scale requires knowledge of interlinked social and technical systems across varied systems of provision, distinct geographies and built environments.

Acknowledgements Thank you to the workshop attendees in Sheffield, January 2024: Linda Clarke, Westminster University; Gavin Killip and Richard Bull, Centre for Sustainable Construction and Retrofit, Nottingham Trent University; Charlotte Surrey, The Green Register; Sandy Rushton, People Powered Retrofit;

Cara Jenkinson, Ashden; Donal Brown, Ashden; Alastair Mumford, MCS; Ceri Batchelder, SYCA, Ellora Coupe, Her Own Space; Jenny Brierley, Breathable Homes; Leah Robson, Your Energy Your Way; Martin Fletcher, Leeds Beckett University; Alice Corovessi, INZEB; Gloria Callinan, TUS; Drogomir Tzanev, ENEffect and Vilislava Ivanova, E3G; and the chapter reviewers.

The contributions of authors were supported by the following funding sources: Hadi Arbabi, Will Eadson, and Rachel Macrorie—South Yorkshire Sustainability Centre (Research England Development Fund); Richard Hanna—UK Energy Research Centre (EP/S029575/1); Kaylen Camacho McCluskey—Research England Policy Support Fund; Kate Simpson, Centre for Sustainable Construction and Retrofit: Strategic Investment Fund, NTU; Faye Wade—UKRI Energy Demand Research Centre (EP/Y010078/1)

References

Bieler, A., Joosse, A., Bano, N., & Jacob, J. (2019). *Final report on OCWI's shared apprenticeship model*. Ontario Centre for Workforce Innovation.

Brocklehurst, F., Morgan, E., Greer, K., Wade, J., & Killip, G. (2022). Domestic retrofit supply chain initiatives and business innovations: An international review. *Buildings & Cities*, 2(1), 533–549.

BPIE. (2024). *Renocally: Empowering municipal renovation action plans and use of technical assistance*. BPIE. https://www.bpie.eu/renocally/

Cauvain, J., Karvonen, A., & Petrova, S. (2018). Market-based low-carbon retrofit in social housing: Insights from Greater Manchester. *Journal of Urban Affairs*, 40(7), 937–951.

CEDEFOP. (2023). *Construction workers: Skills opportunities and challenges (2023 Update)*. CEDEFOP. https://tinyurl.com/3s3hjd3n

CIOB. (2022). *UK crying out for female tradespeople says research by the Chartered Institute of Building*. CIOB. https://www.ciob.org/news/uk-crying-out-for-female-tradespeople-says-research-by-the-chartered-institute-of-building

CITB. (2024). *Equality, diversity and inclusion*. CIOB.

Clarke, L., Michielsens E., Snijders, S., & Wall, C. (2015) *No more softly, softly: Review of women in the construction workforce*. University of Westminster. https://core.ac.uk/download/pdf/161107132.pdf

Clarke, L., Michielsens, E. & Snijders, S. (2017) Misplaced gender diversity policies and practices in the British construction industry: Developing an inclusive and transforming strategy. In F. Emuze & J. Smallwood (Eds.), *Valuing people in construction*. Routledge.

Clarke, L., Sahin-Dikmen, M., & Winch, C. (2020). Transforming vocational education and training for nearly zero-energy building. *Buildings and Cities*, *1*(1), 650–661.

Clegg, S., Loosemore, M., Walker, D., van Marrewijk, A., & Sankaran, S. (2023). Construction cultures: Sources, signs, and solutions of toxicity. In S. Addyman, & H. Smyth (Eds.) *Construction project organising* (pp. 3–17). Wiley-Blackwell.

Commission and Directorate-General for Internal Market, Industry, Entrepreneurship and SMEs, , 2023 European Commission, Directorate-General for Internal Market, Industry, Entrepreneurship and SMEs. (2023). *Transition pathway for construction*. Publications Office of the European Union.

CORDIS. (2023). Leveraging new skills for the building sector to deliver on the European Green Deal. In *CORDIS Results Pack on construction skills: A thematic collection of innovative EU-funded research results* (2nd ed.). European Commission.

Crespo Sánchez E., López Plazas F., Onecha Pérez B., & Marmolejo-Duarte C. (2023). Towards intergenerational transfer to raise awareness about the benefits and co-benefits of energy retrofits in residential buildings. *Buildings*, *13*(9), 2213.

European Construction Sector Observatory. (2020). *Improving the human capital basis—Analytical report*. https://ec.europa.eu/docsroom/documents/41261

European Commission. (2020). *A renovation wave for Europe—Greening our buildings, creating jobs, improving lives*. COM(2020) 662 final. European Commission.

European Commission. (2021). *Questions and answers on the revision of the energy performance of buildings directive, European Commission, 15 December 2021*. European Commission. https://ec.europa.eu/commission/presscorner/detail/en/qanda_21_6686

European Commission. (2023). *Businesses in the construction of buildings sector*. European Commission. https://ec.europa.eu/eurostat/statistics-explained/index.php?oldid=305530#Size_class_analysis

EEA. (2023). *Greenhouse gas emissions from energy use in buildings in Europe*. European Environment Agency (EEA). https://www.eea.europa.eu/en/analysis/indicators/greenhouse-gas-emissions-from-energy?activeAccordion=ecdb3bcf-bbe9-4978-b5cf-0b136399d9f8

European Union. (2023). *Directive (EU) 2023/1791 of the European Parliament and of the Council of 13 September 2023 on energy efficiency and amending regulation (EU) 2023/955 (recast)*. Official Journal of the European Union.

European Union. (2024). *Directive (EU) 2024/1275 of the European Parliament and of the Council of 24 April 2024 on the energy performance of buildings (recast)*. Official Journal of the European Union.

Galvin, R., & Sunikka-Blank, M. (2014). The UK homeowner-retrofitter as an innovator in a socio-technical system. *Energy Policy, 74*, 665–662.

Geonardo. (2024). *European skills registry*. https://skillsregistry.eu/

Green, A. (2016). *Low skill traps in sectors and geographies: Underlying factors and means of escape (September)*. Institute for Employment Research, University of Warwick.

Killip, G. (2020). A reform agenda for UK construction education and practice. *Buildings and Cities, 1*(1), 525–537.

Macrorie, R., Arbabi, H., Camacho McCluskey, K., Eadson, W., Hannah, R., Simpson, K., & Wade, K. (2024). Workshop materials: Output summary notes; participant expertise register. *Zenodo*. https://doi.org/10.5281/zenodo.11202215

Marmolejo-Duarte, C., Arenas, R., Berrio, S., Spairani, Y., Lamas, C. (2022). One-stop-shops as a model to manage housing energy retrofit. A General Approach to Europe and Spain. in: *PLEA 2022: Will Cities Survive? The future of sustainable buildings and urbanism in the age of emergency, 2*, 979–984.

Oktave. (2024). *L'accompagnement Oktave dans la rénovation énergétique*. https://www.oktave.fr/accompagnement-renovation-energetique/

Renovate Europe & E3G. (2022). *Briefing: Speeding up the delivery for renovation—Investing in skills*. Renovate Europe.

Regnier, C., Mathew, P., Robinson, A., Shackleford, J., & Jiron, A. (2023). *System retrofits in efficiency programs: Track record and outlook*. Lawrence Berkeley National Laboratory.

Stroud, D., Antonazzo, L., & Weinel, M. (2024). "Green skills" and the emergent property of "greening." *Policy Studies*, 1–20. https://doi.org/10.1080/01442872.2024.2332441.

Topriska, E., Kolokotroni, M., Melandri, D., McGuiness, S., Ceclan, A., Christoforidis, G.C., Fazio, V., Hadjipanayi, M., Hendrick, P., Kacarska, M., & Peñalvo López, E. (2018). The social, educational, and market scenario for nZEB in Europe. *Buildings, 8*(4), 51.

Wade, F., Shipworth, M., & Hitchings, R. (2016). Influencing the central heating technologies installed in homes: The role of social capital in supply chain networks. *Energy Policy, 95*, 52–60.

Wade, F., Webb, J., & Creamer, E. (2022). Local government capacities to support net zero: Developing comprehensive heat and energy efficiency strategies in Scotland. *Energy Research & Social Science, 89*, 102544.

Open Access This chapter is licensed under the terms of the Creative Commons Attribution 4.0 International License (http://creativecommons.org/licenses/by/4.0/), which permits use, sharing, adaptation, distribution and reproduction in any medium or format, as long as you give appropriate credit to the original author(s) and the source, provide a link to the Creative Commons license and indicate if changes were made.

The images or other third party material in this chapter are included in the chapter's Creative Commons license, unless indicated otherwise in a credit line to the material. If material is not included in the chapter's Creative Commons license and your intended use is not permitted by statutory regulation or exceeds the permitted use, you will need to obtain permission directly from the copyright holder.

PART IV

Navigating the Delivery of New Technology

CHAPTER 7

Promote Integrated Policy Design to Overcome Social and Technical Challenges for Agrivoltaic Deployment

Alessandro Sculio, Pınar Derin-Güre, Ivan Gordon, Angela Ciotola, and Hanna Dittmar

Policy Highlights To achieve the recommendation stated in the chapter title, we propose the following:

- Provide an overarching shared definition of Agrivoltaics (AV) with a precise balance between the agriculture and energy components—general enough to guarantee agriculture will be kept as the primary activity but flexible enough to adapt to the specific conditions.
- Fund further Research and Development (R&D) through pilot projects to investigate different technical configurations and crucial

---

A. Sculio
Department of Cultures, Politics and Society, University of Torino, Torino, Italy
e-mail: alessandro.sciullo@unito.it

P. Derin-Güre (✉)
Middle East Technical University, Ankara, Turkey
e-mail: pderin@metu.edu.tr

© The Author(s) 2024
A. Crowther et al. (eds.), *Strengthening European Energy Policy*,
https://doi.org/10.1007/978-3-031-66481-6_7

parameters to improve the balance between energy production and farming.
- Identify an integrated scheme of incentives for AV to prevent agriculture from being abandoned in favour of energy production.
- Engage with local communities to develop local and national criteria for implementing AV, support the continuation of agricultural activity, raise local awareness, and foster active community involvement in AV projects to maximise regional economic benefits and avoid extractivist behaviours.
- Facilitate the dialogue between Social Sciences and Humanities (SSH) and Science, Technology, Engineering and Maths (STEM) communities by constructing a common definition of the AV policy problems at stake.

Keywords Agrivoltaics · Solar energy · Semi-structured interviews · Socio-technical · Europe

I. Gordon
IMEC, Leuven, Belgium

Hasselt University, Hasselt, Belgium

Department of Electrical Sustainable Energy, Delft University of Technology (TU Delft), Delft, The Netherlands

I. Gordon
e-mail: Ivan.Gordon@imec.be

A. Ciotola
Institute for Technology Assessment and Systems Analysis, Karlsruhe Institute of Technology, Karlsruhe, Germany
e-mail: angela.ciotola@kit.edu

H. Dittmar
SolarPower Europe, Brussels, Belgium
e-mail: h.dittmar@solarpowereurope.org

ETIP PV Secretariat, Brussels, Belgium

7.1 Introduction: Agrivoltaic Deployment as a Socio-Technical Innovation Pathway

This chapter focuses on policy recommendations for the Agrivoltaic (AV) sector, which is considered a pivotal trigger for the energy transition. We adopt a socio-technical approach to consider the interplay among technological, economic, and societal components in deploying Renewable Energy System (RES) technologies. Energy transition has been widely recognised as a paradigmatic socio-technical innovation, involving the co-evolution of technological and societal (cultural, regulatory, and economic) components (Geels & Schot, 2007). The promotion of interdisciplinary dialogue among Technical and Social Sciences has been gaining relevance in identifying complex socio-technical challenges to feed the design and implementation of effective policies tackling this complexity. With this in mind, an interdisciplinary workshop was held in November 2022 in Torino (Italy) to bring together both the Social Sciences and Humanities (SSH) and Science, Technology, Engineering, and Mathematics (STEM) communities belonging to the Joint Programmes e3s and PV of the EERA consortium.[1] The focus was identifying the specific socio-economic and technological challenges for large-scale PV deployment, recognising AV as a promising and challenging socio-technical innovation requiring joint SSH-STEM research.[2]

Solar energy, in general, stands as a foundational element in the EU's journey towards cleaner energy under the European Green Deal and the RePowerEU plan (European Commission, 2019, 2022). However, in densely populated areas (e.g., Western Europe), competition between energy generation and farming to use large plots of land is a crucial issue in energy policy design. AV has excellent potential to address this competition, as it uses land for simultaneous solar energy production and agriculture production. This can reduce the competition between food and energy systems while helping meet energy and food demands (Gomez-Casanovas et al., 2023). Therefore, AV, where surface area is used for multiple purposes, including energy production, are vital for a successful low-carbon energy transition by promoting complementary, rather than competitive, relationships between critical land uses (Pascaris et al., 2021). AV represents a testbed to promote interdisciplinary and cross-sectoral dialogue and policy design.

The deployment of AV is a multifaceted process running on many scales (local, national, global) and various domains (technical, regulatory, cultural) that requires an equally multi-dimensional and integrated perspective in its analysis and the definition of policy-supporting tools (Casanova, 2023). AV deployment involves multiple actors, processes, and interests that might hamper, or even stop, the innovation pathways of AV. The different perspectives between project implementers and local communities regarding the trade-off between (1) land use for farming and energy production, (2) the balance between jobs lost and jobs gained, and (3) the procedural constraints of the authorisation processes are just a few examples of the dynamics that might slow AV adoption. All this clearly illustrates the need for an effective dialogue between STEM and SSH experts aiming to solve this complex matter.

This chapter draws upon a mixed, qualitative research methodology (literature review, questionnaire, interviews) implemented by an interdisciplinary team consisting of experts in STEM (Physicists and Engineers) and SSH (Sociologists, Economists, and Lawyers) aimed at refining what the scientific and policy communities have identified as the main challenges and possible solutions for AV deployment in Europe. A literature review was conducted to get the broader context of existing policies, regulations, and institutional frameworks that impact the deployment of AV. A questionnaire was conducted, with the questionnaire designed jointly by SSH and STEM researchers as it included both technical and social questions. Semi-structured interviews were conducted with various European stakeholders (policymakers, farmer associations, PV Agrivoltaic developers, and researchers) based on their disciplinary background (technical vs non-technical) and their role in the AV field (Sculio et al., 2024), to get an original and critical view of the growing knowledge of AV deployment obstacles and opportunities. Our policy recommendation was developed based on insights obtained through these three methods.

7.2 Challenges for AV Deployment

Based on the literature review, the challenges for AV deployment can be clustered into five main dimensions: Conceptual, Technological, Economic, Social and Environmental, and Institutional. The interview questions were informed by these dimensions. Each cluster is briefly described in this section, and the interview insights are reported.

Regarding the *Conceptual dimension*, a common definition of AV is not universally accepted and adopted in the policy and scientific domains. AV systems must integrate Photovoltaics (PV) with farming activities, which must be kept as the primary goal. The lack of definition leads to varying amounts of farming and electricity production. One interviewee remarked, "*If we only had electricity, I would not consider it Agrivoltaic anymore*" (I4). Another said, "*It is tough to find a definition because agriculture is so broad, but a not perfect standard (SPEC 91434[3]) or definition is much better than no definition*" (I1). This is a crucial societal and policy issue since defining an AV system and prioritising energy production versus farming determines how the supporting policies are defined. An interview participant commented, "*It is essential to separate free field[4] PV and AV. AV is different concerning subsidies, taxes, and the heritage of lands… one thing to prevent is pseudo-agriculture, so people pretend to have agriculture fields, but they only have PV*" (I3). Additionally, the "*EU does not require member states to have definitions in the Common Agriculture Policy*" (I5), so integrating AV into incentive schemes remains unclear.

As for the *technological dimension*, even though research on AV systems has increased, technical challenges must be addressed to maximise electricity generation, while minimising the negative impact on crop yield. Several aspects of this question refer to the trade-offs emerging through the integration with agriculture. On the energy side, "*it is not expected to have the optimum electricity production because PV settings should adapt to agriculture. On the farming side, the impacts depend on the selected crop and several site-specific factors such as the soil, area and climatic conditions*" (I4). Production might be affected in terms of productivity (i.e., the quantity of yield) and organoleptic and nutritional properties (i.e., quality of the yield).

In terms of the *Economic dimension*, there are many, diverse challenges for AV users that can be grouped around four main categories: (1) dealing with the financial aspects of AV related to funding installations, assessing the profitability of the investments, and considering the impact on the quality and quantity of crops; (2) expected but unpredictable changes in the price of land; (3) assessing the extent to which AV can mitigate the competition between energy; and (4) agriculture production impacting the income of the local communities (Al Mamun et al., 2022; Chatzipanagi et al., 2023). Given that the primary activity of farmers should be farming, attention is primarily paid to the impact of AV on farmers' income and farming production, both affected by three central dynamics that can be triggered by AV deployment. Firstly, the increase in the price of agricultural land that can host AV might result in barriers to expanding agriculture activities. Secondly, AV can have different adverse effects on farmers' income depending on the different crops, with greater decreases expected for arable than horticultural crops. Finally, a distortion might quickly arise in connection with the economic benefit, if the net income for farmers adopting AV increases to the point that they could give up the cultivation of crops and focus on electricity production. In addition to hampering the maintenance and profitability of agriculture activity, such dynamics might change the role and identity of farmers from being well-established and recognised actors in the food sector to being new (and underrepresented) actors in the energy sector. Considering the unpredictable socio-economic distortions that might arise from AV deployment, attention should be paid to guaranteeing that AV is always an additional, rather than the primary, source of income for farmers.

As for the *Social and environmental dimensions*,[5] AV deployment might impact the environment, landscape, adopters of AV, recipient communities, biodiversity, and the surrounding ecosystem (Casanova, 2023; Hu, 2023; Taylor et al., 2023). Regarding the effect on the landscape, opinions diverge due to both the objective diversity of AV systems under scrutiny, and the subjective perspective of the observer. It seems true that "*...in some rural landscapes - climate-friendly areas and beautiful landscape - people say that we do not have to put [PV] panels there because the artificialisation of the soil could alter the landscape*" (I5). However, it is also true that "*...the beauty and landscape are perceptions and people see it differently*" (I4), and not all modifications of the natural landscape are harmful.

Unlike other RES technologies, AV should, in principle, enable recipient communities to be more involved, but this potential is not adequately exploited. An optimistic view highlights that while "*local communities do not accept solar parks because big companies usually buy land and are the only ones profiting from it, AV could be different, and communities could directly benefit from energy production either as producers or direct users*" (I3). Yet, for local communities to actively participate in AV, special effort needs to be made to involve farmers and communities, as they "*need to know why the promoters intend to use the system, the benefits, and the trade-offs. We should be honest; it should be their decision to adopt it or not*" (I4). In short, given that public awareness and acceptance among farmers and rural communities are pivotal components influencing AV deployment, attention must be paid not only to which impacts are likely to be produced by AV deployment, but also to how these impacts are perceived by, communicated to, and discussed with the affected communities. Timely and effective engagement strategies should be put in place by the AV promoters.

Community engagement relates to the *Institutional dimension of AV*, which refers to regulations and policies. It has been recognised that involving people, not only helps increase acceptance, but can also play a role in defining effective regulations (Bryner, 2001) - *"farmers and people are critical and should be involved in the regulations. Suppose the farmer wants to install an AV. In that case, many regulations are costly, so they should ask a middle person to facilitate the interpretation of the regulations and find funds. This is why we should go in this direction; government and policies should follow this"* (I3). This uncertainty about the rules that are regulated by experts could be mitigated by involving the final recipients—this is typical of new policy fields such as AV deployment.

Heterogeneity among EU Member States produces a lack of harmonisation of legislation in the EU. In several Member States, land characterisation may change after the realisation of an AV installation, introducing legal obstacles for the farmer. Such a change may result in exclusion from the Common Agricultural Policy (CAP).[6] Moreover, farmers may lose agricultural subsidies. Uncertainties and financial consequences may result in a perception of legal insecurity and possible loss of income for the farmer or the investor. All stakeholders interviewed noted that AV is still poorly incorporated in the Member State's national strategic plans and policies, risking that AV deployment might fail to increase complementarity between energy and agriculture production.

7.3 Achieving Our Recommendation

The Joint Research Council (JRC) report on the potential of and challenges for AV in the European Union, which included 17 possible interventions at the EU level, was used as a stimulus to support interview discussions (see Sculio et al., 2024). Interview participants were asked to reflect upon and prioritise the policy options; in doing so, they highlighted shortcomings of existing policy options, identified additional policy recommendations, discussed implementation challenges, and commented on the interaction between policies. These discussions, coupled with the literature review, supported the identification of our policy recommendation to *promote integrated policy design to overcome social and technical challenges for agrivoltaic deployment*, supported by the following four sub-recommendations.

There is a need *to define AV and implement standardised systems across the EU to ensure harmonisation of policies, develop adequate standards for AV, and differentiate AV systems from traditional PV systems on agricultural land.* The latter is crucial to prevent greenwashing, as is streamlining permitting processes and prioritising grid connections. For this aim, STEM and SSH expertise must be carefully merged to clearly understand the expected balance of energy and agriculture production for a PV system to be considered AV and mitigate the risk of triggering a distortion in the pre-existing agriculture production and culture.

Further Research and Development (R&D) funding, using national and EU level funds, through pilot projects is needed to investigate different technical configurations and establish AV quality standards and parameters to improve the balance between energy production and farming, which could include farmers and researchers from PV and Agriculture. Here, the focus is on promoting real-world experiments that can grasp the interaction among the many site-specific variables in various socio-economic, environmental, and farming contexts. STEM expertise is crucial for designing, implementing, and monitoring projects, while SSH should engage and create a framework for social and economic impacts.

An adequately *integrated scheme of incentives* for AV needs to be established to avoid abandoning agriculture in favour of energy production, with specific attention paid to integrating AV projects into the CAP framework. It is essential to ensure the continuity of farming activities and land preservation post-AV deployment by assuring transfer rights will not be disadvantageous in case of farm inheritance, and to guarantee CAP subsidies for farms with certified AV systems. The promotion of AV through CAP strategic plans, coupled with dedicated financial support and capacity targets at the national level, underscores a commitment to fostering AV adoption. These delicate aspects need SSH expertise, mainly economic, with support from juridical and sociological fields for regulatory integration and assessing inequality.

Local communities must be effectively engaged, as the centrality of farmers and rural communities in AV promotion, economic benefits, and property security are highlighted alongside efforts to enhance public awareness and acceptance of AV initiatives. There are two ways in which communities need to be engaged. The first refers to the need to reinforce *AV deployment strategies by taking advantage of embedded local practical farming knowledge* to develop local and national criteria for implementing AV, and support the continuation of agricultural activity, define capacity targets suitable to the specific contexts, provide effective regulatory framework, and support spatial planning by identifying suitable agricultural land for AV deployment. The second refers directly to the *local awareness-raising and active involvement of the community in AV projects* to maximise benefits for the regional economy and employment, and to avoid predatory behaviours from a few specialised companies promoting AV plants from outside.

The dialogue between STEM and SSH expertise should be well-designed and continuous for the local engagement to be successfully implemented. SSH must put in place all the theoretical and methodological tools to assess local communities' needs and opportunities, and define the most suitable engagement strategies. At the same time, STEM is required to support the design and implementation of AV projects by integrating local knowledge and expertise to promote a co-design approach.

Acknowledgements We thank EERA Joint Programmes e3s, PV, ETIP (The European Technology and Innovation Platform), PV STEM, and SSH participants. We thank Valentina Cavanna for providing legal support regarding the questionnaire. This research is also funded by the EU Project SolarHub: A Greek-Turkish Solar Energy Excellence Hub to Advance the European Green Deal under Grant Agreement number 101086110.

Notes

1. European Energy Research Alliance (EERA) is a wide consortium comprising academia and research institutes from around the EU and is structured in 18 Joint Programmes. The two involved in both the workshop and the chapter are JP clean Energy Transition for Sustainable Society (e3s) and JP Photovoltaic Solar Energy (PV). More details at, https://www.eera-set.eu/.
2. Short proceedings of the Turin workshop are available at https://www.eera-e3s.eu/event/3366:let-the-sunshine-in-addressing-the-challenges-for-pv-exploitation-in-the-eu-17.html.

3. Details on the standard can be found here: https://www.normadoc.com/english/din-spec-91434-2021-05.html.
4. Ground-mounted PV systems, also known as free-field solar power plants.
5. We considered jointly the *Social and environmental dimensions*, as the scope of the chapter is mostly on the local impact of AV that allows to consider better the interaction among the different social and technical dimensions; therefore, we do not concentrate on the mitigating effect of AV in terms of the environment.
6. More information on CAP can be found at the following link: https://eu-cap-network.ec.europa.eu/common-agricultural-policy-overview_en.

REFERENCES

Al Mamun, M. A., Dargusch, P., Wadley D., Zulkarnain, N. A., & Abdul Aziz, A. (2022). A review of research on agrivoltaic systems. *Renewable and Sustainable Energy Reviews, 161.* https://doi.org/10.1016/j.rser.2022.112351

Bryner, G. (2001). Cooperative instruments and policy making: Assessing public participation in US environmental regulation. *European Environment, 11,* 49–60. https://doi.org/10.1002/eet.245

Casanova, N. G. (2023). Knowns, uncertainties, and challenges in agrivoltaics to sustainably intensify energy and food production. *Cell Reports Physical Science, 4*(8). https://doi.org/10.1016/j.xcrp.2023.101518

Chatzipanagi, A., Taylor, N., & Jaeger-Waldau A. (2023). *Overview of the potential and challenges for agri-photovoltaics in the European Union* (JRC Science for Policy Report). JRC. https://doi.org/10.2760/208702

European Commission. (2019). *The European Green Deal.* COM(2019) 640 final. European Commission.

European Commission. (2022). *REPowerEU Plan.* COM(2022) 230 final. European Commission.

Geels, F. W., & Schot, J. (2007). Typology of sociotechnical transition pathways. *Research Policy, 36,* 399–417. https://doi.org/10.1016/j.respol.2007.01.003

Gomez-Casanovas, N., Mwebaze, P., Khanna, M., Branham, B., Time, A., DeLucia, E. H., Bernacchi, C.J., Knapp, A. K., Hoque, M. J., Du, X., Blanc-Betes, E., Barron-Gafford, G. A., Peng, B., Guan, K., Macknick, J., Miao, R., and Miljkovic, N. (2023). Knowns, uncertainties, and challenges in agrivoltaics to sustainably intensify energy and food production. *Cell Reports Physical Science*, 4(8). https://doi.org/10.1016/j.xcrp.2023.101518

Hu, Z. (2023). Towards solar extractivism? A political ecology understanding of the solar energy and agriculture boom in rural China. *Energy Research & Social Science*, 98. https://doi.org/10.1016/j.erss.2023.102988

Pascaris, A. S., Schelly, C., Burnham, L., & Pearce, J. M. (2021). Integrating solar energy with agriculture: Industry perspectives on the market, community, and socio-political dimensions of agrivoltaics. *Energy Research & Social Sciences*, 75, 102023.

Sculio, A., Derin-Güre, P., Gordon, I., Ciotola, A., & Dittmar, H. (2024). *Interviews with agrivoltaics stakeholders*. Zenodo. https://doi.org/10.5281/zenodo.11404132

Taylor, M., Pettit, J., Sekiyama, T., & Sokołowski, M. M. (2023). Justice-driven agrivoltaics: Facilitating agrivoltaics embedded in energy justice. *Renewable and Sustainable Energy Reviews*, 188. https://doi.org/10.1016/j.rser.2023.113815

Open Access This chapter is licensed under the terms of the Creative Commons Attribution 4.0 International License (http://creativecommons.org/licenses/by/4.0/), which permits use, sharing, adaptation, distribution and reproduction in any medium or format, as long as you give appropriate credit to the original author(s) and the source, provide a link to the Creative Commons license and indicate if changes were made.

The images or other third party material in this chapter are included in the chapter's Creative Commons license, unless indicated otherwise in a credit line to the material. If material is not included in the chapter's Creative Commons license and your intended use is not permitted by statutory regulation or exceeds the permitted use, you will need to obtain permission directly from the copyright holder.

CHAPTER 8

Increase Social Acceptability of Nuclear Fusion, Agrivoltaics, and Offshore Wind Through National Support Programmes

Pascal Clain, Insaf Khelladi, Christophe Rodrigues, Alessandro Biancalani, Guillaume Guerard, and Saeedeh Rezaee Vessal

Policy Highlights To achieve the recommendation stated in the chapter title, we propose the following:

- Facilitate the establishment of observatories to monitor social acceptability of low-carbon energy technologies at the EU and national levels.

P. Clain (✉) · I. Khelladi · C. Rodrigues · A. Biancalani · G. Guerard · S. Rezaee Vessal
Léonard de Vinci Pôle Universitaire, Paris, France
e-mail: pascal.clain@devinci.fr

I. Khelladi
e-mail: insaf.khelladi@devinci.fr

C. Rodrigues
e-mail: christophe.rodrigues@devinci.fr

© The Author(s) 2024
A. Crowther et al. (eds.), *Strengthening European Energy Policy*,
https://doi.org/10.1007/978-3-031-66481-6_8

- Offer technical assistance to help Member States incorporate social acceptability factors into their energy transition strategies.
- Develop training programmes to integrate social acceptability factors into the design of low-carbon energy projects from the start.
- Assist countries in managing and resolving disputes and interactions regarding different low-carbon energy technologies.
- Social Sciences and Humanities (SSH) Science, Technology, Engineering and Mathematics (STEM) collaborative recommendations can ensure policies are informed by a nuanced understanding of technical and social structures, making them more practical and widely accepted.

Keywords Low-carbon energy · Topic modelling · Community engagement · Public perception · France

8.1 Introduction

The REPowerEU Plan involves accelerating Europe's clean energy transition (European Commission, 2022). This means speeding up renewables deployment, including issuing permits for project implementation and reducing the time for project roll-out. For instance, it takes up to 9 years to obtain a permit for onshore or offshore wind energy projects and up to 4.5 years for solar photovoltaics (PVs) (European Commission, 2022). To expedite this process, the European Commission (EC) suggests more participatory approaches that involve local and regional authorities, setting up special 'go-to' geographical areas for renewable energy as a top priority, and creating special zones ('regulatory sandboxes') to generate new ideas. These suggestions call for swift and strong actions, but often ignore the local and national opposition to projects, rarely treating social acceptability as crucial in the early design stages.

A. Biancalani
e-mail: alessandro.biancalani@devinci.fr

G. Guerard
e-mail: guillaume.guerard@devinci.fr

S. Rezaee Vessal
e-mail: saeedeh.vessal@devinci.fr

In this chapter, we investigate the social acceptability of three low-carbon energy technologies, focusing on the case of France: offshore wind turbines (using huge fans in the sea to generate electricity from the wind), agrivoltaics (integrating photovoltaic [PV] modules into agricultural land without reducing its productivity), and nuclear fusion (joining small atoms to make a larger one and release energy). Social acceptability means "the consent of the population in a project or decision resulting from collective judgment that the project or decision is superior to known alternatives, including the status quo" (Gendron, 2014, p. 118). Social acceptability of low-carbon energy projects involves multiple stakeholders at different levels, leading to legitimate arrangements and rules that align with each territory's vision and the stakeholders' preferred development model (Fournis & Fortin, 2015).

In 2015, France committed to the Paris Agreement to address climate change, aiming to keep global temperature increases under 2 °C, and ideally below 1.5 °C. This goal requires shifting from fossil fuels (comprising 60% of France's energy production-40% oil, 20% natural gas and < 1% coal) (RTE, 2022) to cleaner energy sources, and cutting energy use in France by 40% by 2050, back to the 1960s' levels. By 2035, renewable energy's share in electricity production in the French energy mix must reach at least 40% (excluding nuclear energy). However, efforts to increase renewable energy have faced problems, such as difficulties in finding suitable land and struggles in balancing energy needs, environmental restoration, and community concerns (CESE, 2022). Scenarios show that France must boost its renewable energy capacity, regardless of nuclear power's role. Moving to renewable energy also means shifting to decentralised power, leading to more energy projects and potential conflicts with a progressively sensitive civil society (Sebi & Bally, 2023). These problems reveal social acceptability concerns that should be considered in renewable and low-carbon energy projects.

Offshore wind turbines and agrivoltaics are mature technologies in France. In 2022, Europe had around 30 GW of offshore wind farms in operation. France has 16 offshore wind farms installed or planned, totalling 8 GW by 2032, but only one farm is currently operational. Agrivoltaics systems are now expanding significantly. In 2022, ADEME (French Agency for the Environment and Energy Management) recorded 167 agrivoltaics projects in France, with a capacity of 1.3 GW. France's Multiannual Energy Programme (PPE) stipulates that solar energy production in 2028 must reach 44 GW, including agrivoltaics (Ministère de la Transition Écologique et Solidaire, 2020). Conversely, nuclear fusion is perceived as a future technology, facing substantial scientific and technical hurdles that need resolution before the technology reaches the market. The interest in these three approaches stems from their dual role in energy production and carbon–neutral technologies, their frequent social controversy, and their intricate governance challenges.

Our policy recommendation is based on the following assumptions: (1) Various factors spread through public discussions and shape the social acceptability of low-carbon energy technologies over time, (2) How social acceptability appears and varies between the local and the national levels, and (3) Sources of social acceptability and unacceptability exist among the three low-carbon energy technologies, challenging the energy mix acceptability.

We, the authors, come from interdisciplinary backgrounds, including four Science, Technology, Engineering and Mathematics (STEM) researchers, with these specialisations: (1) thermal energy storage and wind power, familiar with French energy regulations; (2) complex system modelling, focusing on agrivoltaics in France and its social acceptance; (3) plasma Physics for nuclear fusion; and (4) Computer Science, handling data collection and analysis. Two Social Sciences and Humanities (SSH)/ Management Science researchers contributed their expertise in innovation adoption and social acceptability at individual and institutional levels. The STEM and SSH contributors' synergy is crucial for this project, due to the topic's intrinsic nature. For example, STEM researchers explained the complex technologies behind low-carbon energies, while the SSH researchers identified aspects of these technologies that could cause community apprehension.

We used topic modelling and sentiment analysis of the keywords in press documents (published in 2013–2023, extracted from the Europresse[1] database (Clain et al., 2024)). We first reviewed prior research on the social acceptability of the three low-carbon energy technologies—offshore wind, agrivoltaics and nuclear fusion—in specific European countries, including France. This review helped us to recognise the social, political, technological, and institutional determinants of social acceptability. It is also laid the foundation for the three assumptions underpinning the recommendation (see Sect. 8.1). We then analysed the French national and local daily newspapers using the Europresse database, searching for articles containing predetermined keywords—we obtained a total of 27,422 articles (from January 1, 2013, to December 31, 2023).

Offshore wind accounted for the highest number of articles (27,781), followed by agrivoltaics (4324) and nuclear fusion (1317) (Clain et al., 2024).

8.2 Unpacking the Social Acceptability of Low-Carbon Energy Technologies

8.2.1 Social Acceptability of Low-Carbon Energy Technologies

Social acceptability addresses the key question about implementing energy projects ('what for?') and is considered ahead of the project's decision. In the low-carbon energy technology context, social acceptability goes beyond the Not-In-My-Backyard (NIMBY) effect. It is not limited to characterising opposers and supporters, but strives to reveal how power relations shape these technologies, their deployment, and people's responses (Batel, 2020).

Social acceptability values citizen intelligence, integrating it into the projects. The exchanges among the stakeholders aim to build common learning in order to reach social consensus (Batellier, 2012). Social acceptability is thus a question of shared values and beliefs, referring to a collective evaluation rather than to individual positions. Nevertheless, perceiving community benefits merely as tools to enhance social acceptability overlooks the complexity of acceptability. In fact, social acceptability from communities is an essential condition before planning permission can be granted (Cowell et al., 2011). Low-carbon energy projects need to consider the macro-economic level (making major social agreements that influence development plans and project structures), the meso-political level (making fair rules and decisions to help different strategies work together and solve major disagreements through planned arrangements), and the microsocial level (coordinating among people or groups to make sense of low-carbon energy projects and work together) (Fournis & Fortin, 2015). Overall, social acceptability not only considers the stakeholders' immediate reaction (agree, tolerate or reject) to low-carbon energy projects, but also deals with their values, attitudes, and beliefs regarding technologies, infrastructures, society, and the environment.

We reviewed 21 studies (period: 2013–2023) about 9 EU countries (Austria, Belgium, Denmark, Finland, France, Germany, Hungary, Poland and Spain) and the UK, focusing on the social acceptance and acceptability of the three mentioned low-carbon energy technologies. Our analysis of the key findings reveals three critical areas: (1) *Public perception and attitude* vary widely due to cultural, socio-economic and political contexts, affecting how technologies are accepted, (2) *Policy and governance* are crucial for adopting and effectively implementing these technologies, significantly influencing public reception, and (3) Success and acceptability rely on more than technical aspects; *contextual factors*—local conditions, economic factors, and project visibility—are also pivotal.

Social acceptability differs across European nations shaped by each country's unique circumstances. For example, in Germany and the UK, offshore wind turbines' acceptability is affected by geographical, environmental, and community impacts. Conversely, agrivoltaics in Germany, Belgium, and Denmark is influenced by how well agricultural and energy policies align. Similarly, the reception of nuclear energy and nuclear fusion in countries such as Austria, Finland, Spain, Belgium, and Hungary hinges on a blend of historical, cultural, and political influences, as well as public awareness and education.

8.2.2   Focus on the Social Acceptability of Offshore Wind Turbines, Agrivoltaics, and Nuclear Fusion in France

Social Acceptability Has Gained Significant Attention in Discussions About Large-Scale Energy Projects in France

French people support wind energy but often oppose local wind farms. The Harris Interactive poll showed that 73% of French people positively viewed wind energy (Lévy et al., 2018). The Institut Français d'Opinion Publique (IFOP) survey in 2021 found 77% of the respondents expressed a positive view of wind power (Chasles-Parot & Chatelet, 2021). Additional surveys indicated a generally favourable perception, with figures ranging between 76% (Lévy et al., 2021) and 71% (Bracq & Sliman, 2021). Nevertheless, two-thirds of wind farm projects encounter resistance and administrative hurdles from local groups opposed to nearby installations, often delaying, or halting these projects. Social acceptability is a delicate and time-consuming process to cultivate and, once formed, is difficult to overturn.

Agrivoltaics Social Acceptance Was Taken for Granted by the French Government
The law passed on 10 March 2023 (République Française, 2023) aimed to accelerate agrivoltaics technology roll-out, emphasising its benefits to farmers and anticipating its social acceptance. Recent studies addressed the social acceptability of agrivoltaics due to institutional and governance concerns (Torma & Aschemann-Witzel, 2023). In France, the decree (République Française, 2024) endorsing the law of March 2023 sets a default maximum PV module coverage of 40% on plots for solar farms over 10 MW. Conversely, France's National Research Institute for Agriculture, Food, and Environment (INRAE) has warned that coverage beyond 20% significantly reduces agricultural output, questioning the economic viability of the agrivoltaics installation (Mongenier, 2023). Agrivoltaics deployment often leads to land-use conflicts and uncertainty among local stakeholders (Carrausse & de Sartre, 2023).

Public Opinion on Nuclear Fusion Is Influenced by the Existing Nuclear Fission Landscape
France's nuclear fission infrastructure is significantly larger than the European average. Germany have been reducing their reliance on nuclear energy, initiating a phase-out in 2010 and have already closed six nuclear fission reactors. Italy completely ceased its national nuclear fission production following a 1987 referendum. France's extensive use of centralised nuclear fission, with 46% of its citizens supporting nuclear energy (ASN, 2023), might lead to more positive public attitudes towards nuclear fusion compared to Germany (Jones et al., 2019). In France, public discussions about nuclear fusion are primarily held in Provence, where the International Thermonuclear Experimental Reactor (ITER, 2024)—a global effort involving members such as the EU, Japan, Russia, the USA, China, South Korea, and India—is under construction. Viewed as a path to clean energy, this international commitment has garnered support from a segment of the public that believes in the potential of nuclear energy.

The Focus on Nuclear in Social and Political Systems Slows Non-Nuclear Low-Carbon Energy Adoption in France
Social acceptability poses a key challenge to France's achievement of a 100% renewable energy mix, including offshore wind power—in France, this scenario is even harder to accept than the one where nuclear power makes up 50% of the energy mix (RTE, 2022). The nuclear influence

dominates the discourse on energy decarbonisation in France. A key argument against offshore wind turbines is that this technology is not essential to France's energy strategy. The growth of other renewables struggles with inconsistent government support, highlighting the French Government's preference for nuclear energy (Desvallées & de Sartre, 2023).

8.2.3 Results and Analysis

The rising number of articles on offshore wind in the 2020s coincided with the enactment of laws accelerating public action (République Française, 2020). By 2022 and 2023, 4000 articles were published annually, 80% by regional newspapers. These laws enabled the wind power industry to launch multiple projects along the same coastline, and paved the way for broadening public debate for the 2024 decision-making on maritime and offshore wind energy.

Agrivoltaics coverage grew in the 2020s, peaking in June 2023 due to the law of March 2023. Concerns over crop shading, waste, and chemicals from agrivoltaics emerged. Debates about the coverage rate for agrivoltaics plants (over 10 MW; 40% versus 20%) questioned the economic viability threshold. Other articles focused on cohabiting between energy production and agriculture.

Nuclear fusion witnessed increased media attention, spiking with significant events, including the Germany-Wendelstein 7-X experiment in 2016[2] and US fusion records in 2021. From 2021 to 2023, there was a notable increase in nuclear fusion discussions, likely driven by more private investments in the technology and the impact of Russia's invasion of Ukraine on talks about Europe's energy self-sufficiency.

The next step of the analysis entailed grouping similar words, selecting keywords, and identifying the main topics in the news. We used topic modelling algorithms to explore technology themes (Clain et al., 2024). Finally, the sentiment analysis revealed public feelings about these technologies over time (Clain et al., 2024).

Factors Shaping Social Acceptability of Low-Carbon Energy Technologies Over Time
The initial topic modelling analysis identified the key factors and stakeholders influencing the social acceptability of the analysed low-carbon energy technologies in France (Clain et al., 2024). Offshore wind is

challenged both technologically and socio-politically as diverse stakeholders—including industry experts, policymakers, and local communities—engage in debates about the technology, with this highlighting the need to consider and balance their different perspectives. Agrivoltaics, which combines energy production with agriculture, involves farmers, technology companies, and local communities, focusing on creating economic–environmental synergies and emphasising the need for effective communication. Nuclear fusion concentrates on innovation and international research collaboration, involving global research bodies and governments, particularly in reactor development and funding. The analysis highlights convergences, including innovation, investment, international cooperation, and technological synergies, alongside distinctions such as technological maturity and local impacts, among the social acceptability factors regarding the three low-carbon energy technologies. Overall, there is a need to gauge and enhance public acceptance and adoption of these technologies.

Social acceptability varies significantly across different levels—locally, offshore wind and agrivoltaics occasionally encounter resistance due to environmental and aesthetic concerns, but are broadly supported as renewable sources. Nuclear fusion faces mistrust, largely due to historical nuclear fission accidents and ongoing concerns over nuclear waste management, despite it promising significant long-term environmental benefits with minimal waste. However, the real environmental impact of nuclear fusion has yet to be assessed. Overall, there is a need to ensure that low-carbon energy technologies are designed with community support and understanding in mind.

Sources of Social Acceptability and Unacceptability of the Three Low-Carbon Energy Technologies

Finally, the sentiment analysis sorted the 10 most positive and the 10 most negative articles associated with each low-carbon energy technology (Clain et al., 2024). The analysis shows that social opposition to low-carbon energy projects emerges most strongly during public inquiries, marking a significant difference from more tempered public debates and opinion polls. During these inquiries, local communities become aware of the projects' real and immediate impacts, sparking more intense reactions. The diverse arguments from opponents cover ecological, landscape, economic, and governance issues, as well as challenges related to French energy policy, local job creation, and financial impacts. Although these ideas are intertwined and complex to untangle, the emerging hierarchy of concerns indicates the priorities and sensitivities unique to each community facing different energy projects. Overall, there is a need to equip individuals with the skills to promote and implement low-carbon energy technologies effectively, and to ensure smooth integration and conflict resolution in the energy sector.

8.3 Achieving Our Recommendation

As the EU accelerates its energy transition, the importance of low-carbon energy technologies' social acceptability cannot be overstated. Our analysis highlights the multifaceted nature of social acceptability, influenced by cultural, socio-economic, and political contexts that vary significantly across countries. In France, technologies, such as offshore wind, agrivoltaics, and nuclear fusion, are received differently, based on local perceptions, institutional contexts, and interactions among various stakeholders. This variance underscores the need for policies that are both technically sound, and culturally and socially informed. Thus, we formulated our policy recommendation—*increase social acceptability of nuclear fusion, agrivoltaics, and offshore wind through national support programmes*—which can be achieved through the following actions addressed to the European Commission.

Observatories should be established for low-carbon energy technologies, to allow the systematic analysis and monitoring of social acceptability factors regarding various low-carbon energy technologies across the EU and at the national level. This initiative would involve collecting data, conducting research, and disseminating findings to inform and guide policy and project implementation. These observatories would help diffuse transparent communication campaigns that clearly outline the energy projects' benefits and potential impacts, addressing concerns proactively to foster trust and acceptability.

Dedicated technical assistance should be provided to EU member states to help them integrate social acceptability factors into their energy transition strategies. This support could include offering advisory services, sharing best practices, and facilitating workshops and seminars to build capacity at national and local levels.

Comprehensive training programmes for facilitators and project managers should be developed and implemented. Dealing with low-carbon energy technologies, these programmes should focus on the integration of social acceptability factors from the early stages of project design, ensuring that facilitators are well-equipped to handle community engagement and conflict resolution.

Countries should be assisted in understanding, and effectively managing, the interactions and disputes arising from various low-carbon energy technologies. This support could involve conflict resolution services, mediation between stakeholders, as well as the development of guidelines for managing technological and sectoral overlaps.

These recommendations aim to address the critical elements of social acceptability identified through our research. By focusing on enhancing community engagement, fostering cross-sectoral collaboration, and educating stakeholders, the EU can ensure a smoother transition to a sustainable energy future. These strategies will help mitigate the risk of social resistance and maximise the societal benefits of transitioning to low-carbon energy sources.

Notes

1. https://www.europresse.com/.
2. The world's largest fusion device of the stellarator type.

References

ASN. (2023). *La perception par les français du nucléaire et de son contrôle – baromètre 2022–2023*. Autorité de Sûreté Nucléaire.
Batel, S. (2020). Research on the social acceptance of renewable energy technologies: Past, present and future. *Energy Research & Social Science, 68*, 101544.
Batellier, P. (2012). *Revoir les processus de décision publique: de l'acceptation sociale à l'acceptabilité sociale* (1er). Gaïa Presse.
Bracq, C., & Sliman, G. (2021). *Regard des Français sur l'énergie éolienne* (p. 24). Odoxa.
Carrausse, R., & de Sartre, X. A. (2023). Does agrivoltaism reconcile energy and agriculture? Lessons from a French case study. *Energy, Sustainability and Society, 13*(1), 8.
CESE. (2022). *Acceptabilité des nouvelles infrastructures de transition énergétique: Transition subie, transition choisie?* (Saisine gouvernementale, p. 131) [Avis]. Conseil Economique Social et Environnemental,

CESE. https://www.lecese.fr/travaux-publies/acceptabilite-des-nouvelles-infrastructures-de-transition-energetique-transition-subie-transition-choisie

Chasles-Parot, M., & Chatelet, A. (2021). *L'image de l'énergie éolienne auprès des Français.* IFOP.

Clain, P., Khelladi, I., Rodrigues, C., Biancalani, A., Guérard, G., & Vessal Rezaee, S. (2024). *Consider social acceptability of contentious low-carbon energy technologies* [Data set]. Zenodo. https://doi.org/10.5281/zenodo.11172226

Cowell, R., Bristow, G., & Munday, M. (2011). Acceptance, acceptability and environmental justice: The role of community benefits in wind energy development. *Journal of Environmental Planning and Management, 54*(4), 539–557.

Desvallées, L., & de Sartre, X. A. (2023). In the shadow of nuclear dependency: Competing pathways and the social acceptance of offshore wind energy in France. *Energy Research & Social Science, 98*, 103029.

European Commission. (2022). *REPowerEU.* COM(2022) 230 final. European Commission.

Fournis, Y., & Fortin, M. J. (2015). Une définition territoriale de l'acceptabilité sociale: pièges et défis conceptuels. *VertigO, 15*(3).

Gendron, C. (2014). Penser l'acceptabilité sociale: au-delà de l'intérêt, les valeurs. Communiquer. *Revue de Communication Sociale et Publique* (11), 117–129.

ITER. (2024). *ITER en France.* https://www.iter.org/fr/org/iterinfrance

Jones, C. R., Yardley, S., & Medley, S. (2019). The social acceptance of fusion: Critically examining public perceptions of uranium-based fuel storage for nuclear fusion in Europe. *Energy Research & Social Science, 52*, 192–203.

Lévy, J.-D., Bartoli, P.-H., & Gautier, A. (2018). *Les Français et l'énergie éolienne – vague 1—Comment les Français et les riverains de parcs éoliens la perçoivent-ils?* Harris interactive.

Lévy, J.-D., Bartoli, P.-H., & Gautier, A. (2021). *Les Français et l'énergie éolienne – vague 2—Comment les Français et les riverains de parcs éoliens perçoivent-ils l'énergie éolienne?* Harris interactive.

Ministère de la Transition Écologique et Solidaire. (2020). *La Programmation pluriannuelle de l'énergie (PPE) 2019–2028.* République Française.

Mongenier, L. (2023). *Le décret sur l'agrivoltaïsme va trop loin.* Lafranceagricole. https://www.lafranceagricole.fr/agrivoltaisme/article/860367/le-decret-sur-l-agrivoltaisme-va-trop-loin

République Française. (2020). *LOI n° 2020-1525 du 7 décembre 2020 d'accélération et de simplification de l'action publique.* https://www.legifrance.gouv.fr/jorf/id/JORFTEXT000042619877

République Française. (2023). *LOI n° 2023-175 du 10 mars 2023 relative à l'accélération de la production d'énergies renouvelables.* https://www.legifrance.gouv.fr/dossierlegislatif/JORFDOLE000046329719/

République Française. (2024). *Décret n° 2024-318 du 8 avril 2024 relatif au développement de l'agrivoltaïsme et aux conditions d'implantation des installations photovoltaïques sur des terrains agricoles, naturels ou forestiers.* https://www.legifrance.gouv.fr/jorf/id/JORFTEXT000049386027

RTE. (2022). *Futurs énergétiques 2050.* Réseau terrestre d'électricité.

Sebi, C., & Bally, F. (2023). *Nucléaire, éolien: quelle évolution du discours médiatique en France?* The Conversation. Retrieved on 24 March 2024 from https://theconversation.com/nucleaire-eolien-quelle-evolution-du-discours-mediatique-en-france-208259

Torma, G., & Aschemann-Witzel, J. (2023). Social acceptance of dual land use approaches: Stakeholders' perceptions of the drivers and barriers confronting agrivoltaics diffusion. *Journal of Rural Studies, 97*, 610–625.

Open Access This chapter is licensed under the terms of the Creative Commons Attribution 4.0 International License (http://creativecommons.org/licenses/by/4.0/), which permits use, sharing, adaptation, distribution and reproduction in any medium or format, as long as you give appropriate credit to the original author(s) and the source, provide a link to the Creative Commons license and indicate if changes were made.

The images or other third party material in this chapter are included in the chapter's Creative Commons license, unless indicated otherwise in a credit line to the material. If material is not included in the chapter's Creative Commons license and your intended use is not permitted by statutory regulation or exceeds the permitted use, you will need to obtain permission directly from the copyright holder.

CHAPTER 9

Protect the EU's Digital Energy Infrastructure Against Cyberthreats Through Advanced Technologies, Human Vulnerability Mitigation, and Ethical Practices

Amal Mersni, Aliaksandr Novikau, Marcin Koczan, and Abdulfetah Abdela Shobole

Policy Highlights To achieve the recommendation stated in the chapter title, we propose the following:

- EU Member States must enforce the implementation of a comprehensive, multi-layered security approach to enhance and strengthen the defence of digital energy systems.
- EU Member States must enforce mandatory cybersecurity training programmes to address human vulnerabilities. Energy operators

---

A. Mersni
Department of Engineering, International University of Sarajevo, Ilidža, Bosnia and Herzegovina
e-mail: amersni@ius.edu.ba

© The Author(s) 2024
A. Crowther et al. (eds.), *Strengthening European Energy Policy*,
https://doi.org/10.1007/978-3-031-66481-6_9

must undertake these programmes to ensure adequate education and promote digital hygiene.
- All EU Member States should deploy Artificial Intelligence ethically to improve the sector's cybersecurity while sharing the technology's benefits equitably with all stakeholders.
- Interdisciplinary approaches, combining Engineering and Social Science insights, can inform recommendations to address complex cybersecurity challenges.

Keywords Layered security · Human factor · Cybersecurity training · Predictive AI · Interdisciplinary collaboration

9.1 Introduction

The EU Green Deal targets reducing greenhouse gas emissions by 55% and boosting the share of energy from renewable sources to 45% (European Commission, 2019). The 2022 REPowerEU plan outlines strategies for conserving energy, diversifying supply, and expanding renewable energy utilisation (European Commission, 2022). Together, these initiatives aim to ensure that all EU citizens can access reliable, affordable, and environmentally sustainable energy (European Commission, 2024).

A. Novikau (✉)
Department of Political Science and International Relations, International University of Sarajevo, Ilidža, Bosnia and Herzegovina
e-mail: anovikau@ius.edu.ba

M. Koczan
Institute of International and Security Studies, University of Wroclaw, Wroclaw, Poland
e-mail: marcin.koczan@uwr.edu.pl

A. A. Shobole
Department of Electrical and Electronics Engineering, Istanbul Sabahattin Zaim University, Istanbul, Turkey
e-mail: abdulfetah.shobole@izu.edu.tr

Digitalisation has already contributed much to energy systems, including advanced energy management and distribution through smart grids, i.e., networked power grid control equipment that relies on Information and Communication Technology (ICT) (Eder-Neuhauser et al., 2017). Smart grids support the integration of renewable energy sources and dynamically balance supply and demand, enhancing grid stability and efficiency. Data analytics and Artificial Intelligence (AI) advances add to energy supply management, addressing variability and storage, optimising energy use, forecasting demand, and implementing preventive maintenance (European Commission, 2023).

Yet, energy sector digitalisation also introduces considerable security challenges. Digitalised energy systems are more vulnerable to cyberthreats since the increasing integration of ICT, smart meters, data collectors, and other connected devices expands the potential entry points for cyber attackers, making possible disruption of service or data theft (ECSO, 2018).

The Internet of Things (IoT) is a network of physical devices embedded with sensors, software, and other technologies connecting and exchanging data with other devices and systems over the Internet, whose inclusion in power networks exposes smart grid devices to wired and wireless cyberattacks (Ghiasi et al., 2023).

Additionally, energy systems have traditionally been split into physical infrastructure used in energy generation and transmission (known as Operational Technology, or OT), and Information Technology (IT) used in the administration domain. However, the energy sector's reliance on smart OT systems has expanded exponentially, removing the separation between physical and digital infrastructure. This shift has implications for security since attacks like ransomware start in IT environments but can potentially spread to OT networks.

Therefore, cybersecurity is at the forefront of the EU's efforts to maintain and develop digitally secure and resilient energy systems. Adopting an interdisciplinary approach to address this complex challenge, which has different technical, legal, ethical, and social dimensions, is essential. The interdisciplinary method combines Science, Technology, Engineering and Mathematics (STEM) fields, which design and implement secure technical systems, with insights from Social Sciences and Humanities (SSH) fields, ensuring that these technologies are socially responsible and legally compliant.

Specifically, this collaboration involves Electrical and Electronics Engineering, Telecommunications Systems and Networks, Public Policy, and Security Studies. The first two provide the technical backbone, delivering advanced solutions for secure infrastructure and reliable data communication. Public Policy insights ensure our recommendations comply with current and future regulations, while Security Studies assess the societal and ethical implications of these technologies, so that the policies mitigate threats, enhance public trust, and reflect EU values.

This chapter presents a recommendation to protect the EU's digital energy infrastructure against an increasingly complex cyberthreat landscape. It highlights the necessity of robust cybersecurity strategies, mandated training programmes, and the ethical integration of cutting-edge technologies across Member States. These measures support the European Green Deal's objectives to decarbonise energy systems and enhance energy efficiency at cost-effective rates.

The methodology that informed the development of the policy recommendation included an in-depth analysis of a wide range of EU policy documents, academic sources, and EU organisational reports, alongside guidelines from well-known security vendors. It draws a comprehensive picture of the current situation from legal, practical, and technical perspectives on the path to a securely digitalised, cybersafe EU energy sector.

The chapter is structured as follows. Section 9.2 discusses common cyberthreats and challenges in the digital energy sector. It analyses the existing EU cybersecurity policy and regulatory framework and explains how to implement comprehensive cybersecurity countermeasures. Section 9.3 concludes and sets out our policy recommendation for EU Member States.

9.2 Cybersecurity in the Digital Energy Sector: Challenges, Current Policies, and Recommended Actions

9.2.1 *Common Cyberthreats and Challenges in the Digital Energy Sector*

Recent years have seen a significant increase in cyberattacks on the critical infrastructure of the energy sector (ECSO, 2018; EnergiCERT, 2022), with the attack intensity and frequency hitting a peak in 2022

(Casanovas & Nghiem, 2023). The energy sector's vulnerability is notably concerning, as it experiences 39% of all cyberattacks (*Security*, 2024). Understanding the origins of these attacks is vital to defeating them. External and internal sources pose considerable risks to the ability of energy systems to function reliably and securely.

External threats include complex cyberattacks conducted by various threat actors, namely individuals or groups carrying out malicious activities by exploiting hardware or software system weaknesses to damage their targets (ENISA, 2023). Some state-sponsored groups engage in cyber espionage and sabotage, threatening national energy security, whereas organised cybercriminals execute coordinated attacks to disrupt energy systems or secure financial gain by using ransomware to hold critical infrastructure hostage, demanding payment for its release (EnergiCERT, 2022).

Advanced Persistent Threats (APTs) represent a substantial cyber risk to the EU digital energy sector. These are prolonged, stealthy cyber campaigns conducted by highly skilled actors aiming to steal sensitive data or spy on organisations over extended periods (Chen et al., 2014; ENISA, 2023). APTs targeting the energy sector could infiltrate energy trading platforms to manipulate market prices or gain unauthorised access to proprietary technology in renewable energy systems.

While most cyberthreats originate from external sources, internal factors are also part of the threat landscape. Insiders, including employees, contractors, and interns—with varying levels of trust and privilege—unintentionally open the door for threat actors in approximately one-third of incidents (*Security*, 2024) through deliberate attacks or accidental errors. Such errors include misconfigured security settings, mishandling sensitive information, or failing to update vital software.

Susceptibility to phishing attacks represents another primary vector for internal threats. Phishing involves persuading potential victims to divulge sensitive information through deceptive means, often appearing as communications from legitimate sources—e.g. emails, messages, or websites that impersonate trusted entities—employing scare tactics, or urgent requests to provoke a response by exploiting the trust and access granted to insiders (ENISA, 2023).

Smart grids and Supervisory Control and Data Acquisition (SCADA) systems are typical targets for internal and external cyberattacks, due to their interconnectedness and essential role in modern energy systems. SCADA is a key technological backbone of the energy sector, enabling monitoring, and controlling processes, to efficiently manage power generation and distribution.

Smart grids are vulnerable to various cyberthreats, such as disruption attacks, primarily through Distributed Denial-of-Service (DDoS), temporarily disrupting critical services and impacting grid operations. Additionally, destructive attacks can cause physical damage to infrastructure, necessitating extensive repairs (Eder-Neuhauser et al., 2017). Another attack targeting smart grids is theft, either of service (in this case, energy services (McLaughlin et al., 2010)), or data, such as sensitive information (Eder-Neuhauser et al., 2017).

SCADA systems' ability to monitor and control grid operations makes them particularly susceptible to cyberthreats, including destructive cyberattacks by APTs. In some incidents, APTs reprogramme Programmable Logic Controllers (PLCs) to alter the functionality of fundamental equipment, misleading operators with false 'normal' operating conditions. Such tactics pose risks of operational disruption and highlight the severe potential for lasting damage to the physical components of the energy sector's infrastructure (Demertzis & Iliadis, 2018).

EU energy sector challenges extend beyond the previously mentioned cyberattacks to broader concerns, encompassing the sector's struggle to ensure cross-border grid stability, adapt to evolving cyberthreats, and integrate cutting-edge technologies (EECSP, 2017). Key issues include standardising cybersecurity measures, protecting key operators, managing supply chain risks, and developing crisis response mechanisms across EU Member States (SGTF EG2, 2019).

Another threat to energy systems is advanced malware fuelled by AI, as it may be highly targeted, activating only under certain conditions, making it harder to detect and more harmful. Thus, traditional cybersecurity approaches need updating to combat this threat. Defence efforts to establish the source of certain malware varieties are complicated because they can blend seamlessly with legitimate software (Blauth et al., 2022).

Given the dynamic nature of cyberthreats, with human errors exacerbating the severity of the risk, and the emergence of AI-powered malware, the EU is well aware that its cross-border energy sector must address present challenges and proactively anticipate forthcoming threats. Strengthening EU defences via standardisation, training, and innovation is not just a suggestion but an absolute necessity to protect the future of EU energy.

9.2.2 Analysis of Existing EU Cybersecurity Policy and Legislation in the Digital Energy Sector

The EU has responded with detailed legislation in the face of the growing cybersecurity threat to crucial energy infrastructure. The Network and Information Security (NIS) Directive (European Union, 2016) was the first piece of EU-wide legislation to attain a uniformly high level of cybersecurity across the Member States.

The NIS2 Directive (European Union, 2022), following NIS, focused on providing the EU energy sector with solid foundations for long-term cybersecurity measures. It elaborated on the importance and effectiveness of detailed and immediate information sharing about incidents. The newly adopted EU Network Code (European Commission, Directorate-General for Energy, 2024) for the electricity sector aims to address energy cybersecurity across the board, on the level of the Union, Member States, regions, and entities.

To secure the digital energy sector, these are highly ambitious initiatives (Table 9.1) for well-integrated and resilient energy production, distribution, management, and maintenance across the EU. However, there are challenges on the path to their realisation. The key limitation is the complicated nature of the EU constitution and institutional structure because Member States and their various regions and entities have varying degrees of competencies, financial means, and instruments.

Traditional cybersecurity measures cannot keep up with the cyberthreat landscape, especially due to AI advancements, which evolves much faster than policy and training efforts at the national level. This creates a security gap, especially for developing economies and lower-income sections of the population, necessitating the EU to provide technical and financial support to ensure a union-wide cybersecure energy supply.

While the newly adopted Network Code provides financing, it is too early to confirm if actual delivery can match the ambition. The first reading suggests the provisions might prove too complicated for the sector's weaker actors to benefit from the available financing.

Despite the comprehensive EU policies, gaps in how modern security approaches are integrated and emphasised still exist. Proposing new guidelines for Member States to adopt these security approaches within their national cybersecurity strategies for the energy sector does not eliminate potential challenges, such as legacy systems, interoperability issues, or the need for sector-specific guidance on how to implement these models.

9.2.3 Implementing Comprehensive Cybersecurity Countermeasures in the Digital Energy Sector

To enhance the cybersecurity of ICT energy systems, many countermeasures, such as technological, educational, and administrative ones, must be implemented, at different levels. Regarding technology, cybersecurity relies on many layers of protection, including perimeter defence, network security, endpoint protection, and application security (McNab, 2017).

Cybersecurity in the constantly evolving threat landscape requires an approach responsive to these evolutions, including shifting from network segmentation to more micro-segmentation (breaking down security perimeters into smaller zones, each requiring different access permission), adopting zero-trust models (multiple security levels, including identity verification, device authentication, application-level security restrictions, and data encryption to ensure that security is not based only on a single point of defence by requiring verification at each layer) over trusted perimeters, shifting focus from threat prevention to response automation (automation of cyberattack and security incident prevention), and expanding protection from networks to assets, data, and digital identities (Cisco, 2023; McNab, 2017; Mukherjee, 2020).

Security in all its forms starts with the human factor, and cybersecurity is no different. The human factor is the one element that could immensely lower the incidence rates and the response time to incidental/accidental cybersecurity breaches. Educating the staff working in and with digitalised energy systems is crucial for overall cybersecurity.

The European Commission has underscored the gap in trained personnel and the inadequacy of current educational curricula to defeat the industry's cyberthreats. The EU Member States have not implemented a comprehensive action plan to mandate staff training to a unified standard, nor created a collective EU-wide educational programme for employees in the energy sector.

Cyberspace is an ever-evolving threat environment, and a minimum level of knowledge is necessary to maintain a safe environment. Thus, training courses following the regular assessment of employee knowledge of the current threat landscape, and different training programmes

Table 9.1 Analysis of key EU cybersecurity legislation specifically for the digital energy sector

EU cybersecurity legislation	Key points	Key challenges
NIS Directive (2016/1148)	• Elevating the network and information systems security across the EU • Establishing national capacities in cybersecurity • Establishing cross-border response to incidents • Improving EU cross-sector cooperation and cyber-resilience • Creating a Computer Security Incident Response Team network to ensure prompt and appropriate collaboration between Member States	• Surge of threat vectors internally and externally • Implementing the NIS Directive at different degrees across the EU • Low cyber-resilience of EU businesses • Lack of joint situational awareness and crisis response
Cybersecurity Act (2019/881)	• Bolstering EU responsiveness to cyberthreats • Improving cyber-resilience and increasing trust in the digital single market • Creating an EU cybersecurity certification model for ICT products, processes, and services • Securing legacy infrastructure	• Hike in ransomware and malware attacks on energy infrastructure • Building cooperation among stakeholders • Uneven responses to the certification requirement across sectors and Member States • Stakeholder willingness to cooperate in updating legacy systems and adapting to the new threat landscape

(continued)

Table 9.1 (continued)

EU cybersecurity legislation	Key points	Key challenges
NIS 2 Directive (2022/2555)	• Increasing the cybersecurity framework to new sectors (such as energy) and entities • Strengthening resilience and responsiveness • Improving preparedness and cooperation among Member States • Establishment of the European Cyber Crises Liaison Organisation Network for large-scale crisis management • Introduction of non-compliance fines on entities	• Financial costs of achieving cross-sector and cross-border cooperation • Stakeholder resistance to implementation • Lack of mechanisms and human resources for implementation in the required time frame
Network Code on Cybersecurity for the Electricity Sector (C/2024/1366)	• Developing EU electricity sector rules • Addressing the cybersecurity aspects of cross-border electricity flows • Cybersecurity risk assessment • Common minimum cybersecurity requirements • Planning, reporting, and monitoring • Crisis management	• Financial costs of upgrading the cybersecurity infrastructure • The cross-linkages with other EU laws to support the outcomes • Lack of human resources to meet the needs of Member States and entities • Lack of coordination across the EU and with third parties outside the EU

for staff from specific departments whose security levels and needs may diverge from each other, are beneficial in addressing human vulnerabilities, mitigating accidental insider threats, and reducing cyber breaches. A cybersecure workplace also requires a company culture, whereby cybersecurity policies are communicated to all staff, and consolidated by regular awareness campaigns, so that cybersecurity becomes part of the company's day-to-day operations at all levels.

The use of AI in energy systems is a subject of ongoing debates (Sovrano & Masetti, 2022) that has been addressed in the EU AI

Act (European Parliament, 2024). However, AI-powered cybersecurity systems can provide proactive threat intelligence by constantly discovering new cyberthreats and responding accordingly. The energy sector may use predictive and data-driven approaches to defend against existing cyberthreats and prepare for future difficulties, necessitating coordination among industry experts, cybersecurity professionals, and legislators to ensure successful, realistic solutions that meet the specific needs of the energy sector. AI examines enormous data volumes to detect potential risks and vulnerabilities before exploiting them. Thus, the use of AI should comply with data protection laws. Energy companies must prioritise data protection and foster a security-focused mindset within their organisations.

9.3 Achieving Our Recommendation

To sustain the energy supply, EU policies and strategies must respond swiftly to the evolving threat landscape. By combining technical and social science expertise, we believe a more holistic understanding of the challenges would bring novel and practical solutions. This interdisciplinary approach informed our policy recommendation, as outlined in our title to *protect the EU's digital energy infrastructure against cyberthreats through advanced technologies, human vulnerability mitigation, and ethical practices*. Our recommendation, and the actions to support its achievement, is applicable across the EU and complements current energy policies.

Enhance and Strengthen Defensive Techniques by Establishing a Comprehensive, Multi-Layered Security Approach to Protect Digital Energy Systems. Cyberthreats are shifting rapidly in scope and structure, and the ICT components in use by the energy sector must, therefore, be able to adapt as swiftly. The strategy to ensure robust cybersecurity must include a layered defence structure that can coordinate to withstand attacks. The layered response should involve an EU-wide stance beyond cross-border information and best practices sharing between Member States. Energy operators must also navigate their pivotal role in protecting infrastructure and collaborate effectively with various suppliers, ensuring clear communication between operators and suppliers across the sector.

Mandate Cybersecurity Training and Professional Development for All Employees. As the human factor is the weakest link in the security chain, all employees must be provided with a set level of compulsory training, followed by professional development, to close the security loop against both external attacks and internal accidents caused by human errors. Therefore, EU-level mandated cybersecurity training plans and programmes for all involvement levels are needed. Energy operators must also enforce these educational and training programmes within their workforce.

Ethically Use Artificial Intelligence (AI) for Predictive Threat Identification to Strengthen and Improve the Energy Sector's Cybersecurity Posture. Proactive response is the most significant benefit of AI-driven systems, enabled by learning-driven defence adaptations. These benefits must be shared equitably across the energy sector actors, from large to small-scale companies sharing know-how and information. Another aspect of ethical employment of AI is developing sector-based ethical frameworks and policies for fair, transparent, and responsible use of AI.

References

Blauth, T. F., Gstrein, O. J., & Zwitter, A. (2022). Artificial intelligence crime: An overview of malicious use and abuse of AI. *IEEE Access, 10*, 77110–77122. https://doi.org/10.1109/ACCESS.2022.3191790

Casanovas, M., & Nghiem, A. (2023). *Current cyberattack trends pose an unprecedented threat to critical infrastructure, such as electricity systems.* IEA. https://www.iea.org/commentaries/cybersecurity-is-the-power-system-lagging-behind

Chen, P., Desmet, L., & Huygens, C. (2014). A study on advanced persistent threats. In C. Salinesi, M. C. Norrie, & Ó. Pastor (Eds.), *Advanced information systems engineering* (Vol. 7908, pp. 63–72). Springer. https://doi.org/10.1007/978-3-662-44885-4_5

Cisco. (2023). *Digitalizing Europe's energy system to power the green energy revolution.* https://storage.googleapis.com/blogs-images/ciscoblogs/1/2023/09/Cisco-Paper-DigitalisationOfEnergy_2023-ClimateWeek.pdf

Demertzis, K., & Iliadis, L. (2018). A computational intelligence system identifying cyber-attacks on smart energy grids. In N. J. Daras & T. M. Rassias (Eds.), *Modern discrete mathematics and analysis* (Vol. 131, pp. 97–116). Springer. https://doi.org/10.1007/978-3-319-74325-7_5

ECSO (European Cyber Security Organisation). (2018). *Energy networks and smart grids: Cybersecurity for the energy sector.* ESCO. https://ecs-org.eu/?publications=https-ecs-org-eu-documents-publications-5fdb2673903c6-pdf

Eder-Neuhauser, P., Zseby, T., Fabini, J., & Vormayr, G. (2017). Cyber attack models for smart grid environments. *Sustainable Energy, Grids and Networks, 12*, 10–29. https://doi.org/10.1016/j.segan.2017.08.002

EECSP. (2017). *Cyber security in the energy sector: Recommendations for the European Commission on a European strategic framework and potential future legislative acts for the energy sector*. Energy Expert Cyber Security Platform.

EnergiCERT. (2022). *Cyber attacks against European energy & utility companies*. https://sektorcert.dk/wp-content/uploads/2022/09/Attacks-against-European-energy-and-utility-companies-2020-09-05-v3.pdf

ENISA (European Union Agency for Cybersecurity). (2023). *ENISA threat landscape 2023: July 2022 to June 2023*. EINSA. https://www.enisa.europa.eu/publications/enisa-threat-landscape-2023

European Commission. (2019). *The European Green Deal*. COM(2019) 640 final. European Commission.

European Commission. (2022). *REPowerEU*. COM(2022) 230 final. European Commission.

European Commission. (2023). *Grids, the missing link—An EU action plan for grids*. COM(2023) 757 final. European Commission.

European Commission. (2024). *Digitalisation of the European Energy System*. https://digital-strategy.ec.europa.eu/en/policies/digitalisation-energy

European Commission, Directorate-General for Energy (2024). *Commission Delegated Regulation (EU) 2024/1366 of 11 March 2024 supplementing Regulation (EU) 2019/943 of the European Parliament and of the Council by establishing a network code on sector-specific rules for cybersecurity aspects of cross-border electricity flows*. Official Journal of the European Union.

European Parliament. (2024). *Artificial Intelligence Act*. TA(2024)0138. European Parliament.

European Union. (2016). *Directive (EU) 2016/1148 of the European Parliament and of the Council of 6 July 2016 concerning measures for a high common level of security of network and information systems across the Union*. Official Journal of the European Union.

European Union. (2022). *Directive (EU) 2022/2555 of the European Parliament and of the Council of 14 December 2022 on measures for a high common level of cybersecurity across the Union, amending Regulation (EU) No 910/2014 and Directive (EU) 2018/1972, and repealing Directive (EU) 2016/1148 (NIS 2 Directive)*. Official Journal of the European Union.

Ghiasi, M., Niknam, T., Wang, Z., Mehrandezh, M., Dehghani, M., & Ghadimi, N. (2023). A comprehensive review of cyber-attacks and defense mechanisms for improving security in smart grid energy systems: Past, present and future. *Electric Power Systems Research, 215*, 108975. https://doi.org/10.1016/j.epsr.2022.108975

McLaughlin, S., Podkuiko, D., & McDaniel, P. (2010). Energy theft in the advanced metering infrastructure. In E. Rome & R. Bloomfield (Eds.), *Critical information infrastructures security* (Vol. 6027, pp. 176–187). Springer. https://doi.org/10.1007/978-3-642-14379-3_15

McNab, C. (2017). *Network security assessment: Know your network* (3rd ed.). O'Reilly Media.

Mukherjee, A. (2020). *Network security strategies: Protect your network and enterprise against advanced cybersecurity attacks and threats*. Packt Publishing.

Security. (2024). Energy sector faces 39% of critical infrastructure attacks. *Security Magazine*. https://www.securitymagazine.com/articles/99915-energy-sector-faces-39-of-critical-infrastructure-attacks

SGTF EG2 (Smart Grid Task Force Expert Group 2). (2019). *Recommendations to the European Commission for the implementation of sector-specific rules for cybersecurity aspects of cross-border electricity flows, on common minimum requirements, planning, monitoring, reporting and crisis management*.

Sovrano, F., & Masetti, G. (2022). *Foreseeing the impact of the proposed AI Act on the sustainability and safety of critical infrastructures*. 15th International Conference on Theory and Practice of Electronic Governance, pp. 492–498. https://doi.org/10.1145/3560107

Open Access This chapter is licensed under the terms of the Creative Commons Attribution 4.0 International License (http://creativecommons.org/licenses/by/4.0/), which permits use, sharing, adaptation, distribution and reproduction in any medium or format, as long as you give appropriate credit to the original author(s) and the source, provide a link to the Creative Commons license and indicate if changes were made.

The images or other third party material in this chapter are included in the chapter's Creative Commons license, unless indicated otherwise in a credit line to the material. If material is not included in the chapter's Creative Commons license and your intended use is not permitted by statutory regulation or exceeds the permitted use, you will need to obtain permission directly from the copyright holder.

PART V

Navigating Models for Policy Development

CHAPTER 10

Understand Stakeholder Perceptions and Implementation Possibilities for Energy Efficiency Measures and Policy Through Multicriteria Modelling

Alexandra Buylova, Aron Larsson, Naghmeh Nasiritousi, and Afzal S. Siddiqui

Policy Highlights To achieve the recommendation stated in the chapter title, we propose the following:

- Stakeholders can be better engaged in energy efficiency decisions through the use of multicriteria models.
- Decision-makers should present trade-offs, such as cost and emissions, and combinations of acceptable solutions to various stakeholders such as the public, housing associations, regulatory agencies, and financial institutions.

A. Buylova (✉) · N. Nasiritousi
Swedish Institute of International Affairs, Stockholm, Sweden
e-mail: alexandra.buylova@ui.se

N. Nasiritousi
e-mail: naghmeh.nasiritousi@liu.se

© The Author(s) 2024
A. Crowther et al. (eds.), *Strengthening European Energy Policy*,
https://doi.org/10.1007/978-3-031-66481-6_10

- Decision-makers should adopt a user-centred approach to energy efficiency measures by encouraging stakeholder dialogues around decision-support tools (e.g. multicriteria modelling) to improve understanding of costs and benefits of measures.
- Decision-makers should identify opportunities for consensus building and mindset shifts about the wider benefits of energy efficiency measures by emphasising their social considerations.
- Using Social Sciences and Humanities (SSH) perspectives can strengthen Science, Technology, Engineering and Mathematics (STEM) led multicriteria models that visualise trade-offs as well as identify plausible conflicts among stakeholders.

Keywords Energy efficiency · Stakeholder engagement · Multicriteria approach · Conflicting objectives · Compromise solutions

---

A. Larsson · A. S. Siddiqui
Department of Computer and Systems Sciences, Stockholm University, Stockholm, Sweden
e-mail: aron@dsv.su.se; aron.larsson@miun.se

A. S. Siddiqui
e-mail: asiddiq@dsv.su.se

A. Larsson
Risk and Crisis Research Centre, Mid Sweden University, Sundsvall, Sweden

N. Nasiritousi
Department of Thematic Studies, Linköping University, Linköping, Sweden

A. S. Siddiqui
Department of Mathematics and Systems Analysis, Aalto University, Espoo, Finland

10.1 Introduction

This chapter assesses trade-offs in urban energy use to advance sustainable planning of cities, contributing to the European Union's (EU) climate-neutrality ambition and the EU Cities Mission. While scientific consensus is clear about climate change, devising a more sustainable pathway for energy consumption is not straightforward. For example, an economy that uses fewer resources and can handle challenges better is a central goal of the European Green Deal (European Commission, 2019). But many people today worry about the cost of living and are concerned about equity, and some political groups use these worries to reject environmental and climate policies (Huber et al., 2021). Thus, the framing of environmental measures as costly interventions that overwhelm peoples' lives makes it hard to make the necessary changes. Moreover, environmental and climate efforts are often portrayed mainly in terms of technological change with limited involvement of users and other stakeholders in the process of planning. Indeed, such an approach omits the possibility to engage the public in shifting to a less resource-intensive energy use (Pineau, 2023). This risks making subsequent policies less acceptable.

In this chapter, we reframe the debate to facilitate stakeholder consensus. We limit the scope of our analysis to buildings, which comprise approximately 40% of EU energy consumption and 36% of energy-related GHG emissions (European Union, 2024). The EU has created numerous directives to improve energy efficiency in buildings, e.g. the revised Energy Performance of Buildings Directive (European Union, 2024) and the Energy Efficiency Directive (European Union, 2023). The former aims to increase the rate of renovations and decrease emissions through measures such as the introduction of renovation passports, targeted financing, and attention to energy poverty, as well as National Building Renovation Plans (European Union, 2024). Moreover, other measures include "one-stop shops for the energy renovations of buildings for homeowners, small and medium-sized enterprises and other stakeholders" (European Union, 2024, p. 38). In this context, stakeholders may include real estate developers, construction firms, landlords, urban planners, funding institutions, environmental agencies, and housing associations. Thus, there exists an opportunity to engage a broad range of stakeholders in energy efficiency decisions to facilitate decarbonisation.

While the regulatory framework aims to enhance buildings' energy efficiency, achieving the intended objectives depends on successful implementation by EU Member States. The two largest barriers to accelerating implementation are lack of financing and lack of knowledge about energy efficiency measures (Carlander & Thollander, 2023; Yeatts et al., 2017). We argue that other barriers include limited engagement of users in decision-making and the framing of possible solutions mainly in terms of costs, which limits the range of acceptable measures. We, therefore, present an approach to facilitate the understanding of how the implementation process can be adjusted to include a range of solutions for improving energy efficiency that not only are cost-effective but also consider CO_2 emissions and other aspects of social wellbeing, while encouraging stakeholder dialogues around decision-support tools (e.g. multicriteria models) to improve understanding of costs and benefits of various measures. The resulting transparency of the approach in illustrating trade-offs could enable consensus around acceptable solutions and uncover incentive-compatible mechanisms for their implementation.

Using STEM expertise, we developed a multicriteria model that visualises trade-offs for different energy efficiency measures in buildings, which was informed by SSH perspectives from Political Science. Intended users of the model are policymakers, but other users (including housing associations, regulatory agencies, and investors) are also targeted. SSH researchers posited plausible conflicts among stakeholders based on previous literature on energy efficiency in buildings, with STEM researchers then devising a quantitative framework for analysis of these conflicts and ways of reconciliation, which was subsequently exemplified for a case of a housing association. A housing association was used as an empirical example, but the model can be adjusted and applied to different settings and scales. A participatory stakeholder workshop was organised to receive feedback on the model. This feedback can further inform model refinement and allow the model to be applicable to different contexts and users.

A literature review showed that European regulation has largely relied on soft measures that overlook direct regulation, incentives, and awareness of the wider benefits of energy efficiency measures in buildings (Carlander & Thollander, 2023; Yeatts et al., 2017). The literature has also identified missed opportunities in cases where building renovations take place without attention to energy efficiency measures (Mjörnell et al., 2019). Limited knowledge about possible energy efficiency measures' costs and benefits is one important barrier. While there are various tools that have been developed to address this drawback, e.g. artificial intelligence (AI) (von Platten et al., 2020), such methods are difficult for users to grasp transparently. Instead, multicriteria analysis has proven effective at communicating attractive solutions and options in real-world settings involving public authorities (Montibeller & Franco, 2011; Phillips, 2011). Thus, this project developed a multicriteria tool to visualise opportunities and trade-offs in energy efficiency measures aimed at user understanding, including policymakers.

The stakeholder workshop tested the model's usefulness as a mechanism for gaining an understanding of options for implementing energy efficiency measures in the built environment. The workshop took place on 11 December 2023 in Stockholm, Sweden and included 12 representatives from public authorities, the building sector, the finance sector, tenants' associations, and researchers working on energy efficiency measures in buildings. Our motivation to engage these stakeholders stems from the fact that these users can be impacted by decision-making regarding energy efficiency policies, but they are often not adequately consulted in the process, which diminishes the effectiveness of policies (Seibicke, 2024). We argue that a deeper understanding of users' needs, and acceptance of policies can be gained through employing a multicriteria model-supported process for participatory governance. Multicriteria models can serve as a tool for organising stakeholder engagement processes and a framework for decision-making. This approach could be applied by the revised Energy Performance of Buildings Directive's one-stop shops for the energy renovations of buildings to encourage stakeholder participation (European Union, 2024).

10.2 Application of a Multicriteria Model for the Built Environment

10.2.1 Handling Conflicting Objectives

The purpose of using a multicriteria model is to foster a transparent assessment of trade-offs between conflicting objectives (Winston & Albright, 2018). We take this perspective to examine the trade-offs between economic and environmental objectives in the context of a building energy management system (BEMS). Our multicriteria model yields a range of intermediate solutions that address conflicting objectives, such as minimising either cost[1] or GHG emissions.[2] In other words, the model provides solutions along a Pareto frontier[3] between cost and GHG emissions minimisation. More importantly, the impacts of energy efficiency measures, subsidies, tariff structures, and technology availability in catalysing changes to building configuration and operations can also be assessed.

Our analysis uses data that reflect an urban housing association in Stockholm, Sweden (Siddiqui, 2024). In our default scenario, the association purchases all energy from suppliers at given prices without any effort to curb consumption. This do-nothing (DN) scenario contrasts with a Base scenario, in which the association may adopt demand- and supply-side measures to reduce consumption, e.g. by renovating the building envelope or installing an on-site generation and storage technologies. Moreover, the association can prioritise either cost minimisation, GHG emissions minimisation, or an intermediate objective between these two extremes. Consequently, the resulting Pareto frontier in the Base scenario (Fig. 10.1) stretches from point A to point B in minimising either cost or GHG emissions, respectively. Meanwhile, points such as C, D, and E identify intermediate solutions that minimise cost for any given level of GHG emissions. Hence, any solution on the Pareto frontier can be interpreted as one in which it is impossible to improve upon one objective without deteriorating the other.

Via this framework, we examine plausible compromises on the Pareto frontier and how the Pareto frontier is affected by energy efficiency measures, energy tariffs, and technology availability. The model shows that *modest* adjustments to existing consumption patterns could reduce GHG emissions. For example, a move from points A to C in the Base scenario leads to a roughly 8% decrease in GHG emissions at a cost increase of 1.5%. Such a transition is facilitated by increasing energy purchases during off-peak hours (through e.g. better utilisation of existing heat storage) to then avoid purchases during peak hours, when more-polluting plants are dispatched. Yet, the scope for this temporal

---

[1] The sum of daily energy purchases, demand charges, amortised cost of efficiency measures, amortised cost of production technologies, and amortised cost of storage technologies.

[2] Emissions from energy purchases since the candidate production and storage technologies for adoption are assumed to be carbon free in operations.

[3] For any given GHG emissions level, this construct indicates the minimised cost.

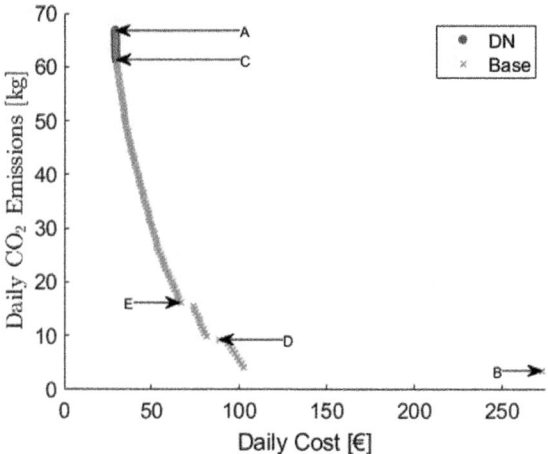

Fig. 10.1 Pareto frontiers that illustrate the trade-off between conflicting objectives for do-nothing (DN) and base scenarios

shift in consumption is limited by the size of the storage unit, which is why the Pareto frontier for the DN scenario extends only from points A to C. In effect, deeper decarbonisation in moving from points C to E requires a larger storage unit to offset GHG emissions from peak-hour heat purchases. Such an investment reduces GHG emissions by 75% relative to those at point A, albeit at more than double the cost. In a similar vein, more ambitious decarbonisation efforts, e.g. to points D and B, necessitate more expansive technology adoption, e.g. building renovations and solar-thermal units. Hence, while points A and B are prominent on the Pareto frontier, intermediate solutions, such as points D and E, indicate the diversity of plausible solutions that could provide compromises around which consensus could coalesce. One of the main points that Fig. 10.1 communicates is that the choices available to stakeholders are much wider than generally perceived and that investments in certain energy efficiency measures can reduce GHG emissions at a much higher rate than the associated cost increases.

The results further indicate that even compromise solutions such as points D and E render a more complicated nexus of energy flows in the exemplar housing association. As a result, an adequate policy framework that supports energy efficiency measures is paramount. For instance, removing the demand charge on the maximum rate of consumption from the energy tariff would increase GHG emissions. Furthermore, energy efficiency measures catalyse more effective use of adopted technologies. Indeed, in tracing out Pareto frontiers, we assume that cost-minimising housing associations could put greater weight on GHG emissions—this would require the provision of, for example, CO_2 taxes, energy tariffs with demand charges, and subsidies on technologies. Thus, policymakers could use multicriteria models to argue the need for, and justify the adoption of, policy support measures, such as taxes and regulations.

10.2.2 Stakeholder Feedback

The practical challenges to implementing such a multicriteria framework were the focus of our participatory stakeholder workshop. Through this forum, we elicited participants' reflections on (1) barriers associated with achieving greater energy efficiency in buildings generally, (2) multicriteria model design and applicability, and (3) regulatory support for the implementation of the framework.

Barriers to Greater Energy Efficiency in Buildings
Stakeholders mentioned broader processes in the energy sector, economy, and policy that affect possibilities for improving energy efficiency in buildings. For example, in the context of Sweden, energy prices are not high enough to incentivise reduction of energy use and investments in efficiency measures, while the volatility of energy prices adversely affects the building sector. Stakeholders also suggested that the tax system needs to be adapted, e.g. tax deductions on home renovations should incentivise energy efficiency measures. Others mentioned the importance of policy packages instead of single measures to prevent one policy from undermining another, for example, if investments into energy efficiency take away investments from other programmes. In addition, energy security and energy poverty were mentioned as important aspects to be considered in the process of policy design. Through the discussions, it highlighted the need to engage a range of stakeholders to weigh-up differing perspectives and reach agreement on ways forward.

Multicriteria Model Applicability and Design
Many workshop participants reflected on the multicriteria model's complexity and their inexperience using it. However, they also saw the

'mindset' of these models as a useful device for understanding trade-offs embedded in the decision-making process. As such, the model itself could be adjusted for ease of use and accessibility to be more widely deployed. On a more conceptual level, it was seen as a tool for scenario analysis to set targets, make trade-offs visible, highlight a business case, use during stakeholder-engagement discussions, and incorporate in building-sector roadmaps. The main feedback from stakeholders about the model design related to barriers, including social and cultural indicators, such as accessibility, safety, wellbeing, and other environmental impacts not being captured. The challenge lies not only in the operationalisation of some of these factors but also in the difficulty of interpreting a complex model. Thus, the model needs to be used alongside other mechanisms and considerations. Yet, the central function of the model remains useful, i.e., demonstrating multiple possibilities along the Pareto frontier, regardless of which attributes are considered. Indeed, as one participant reflected, "*politicians and authorities should require that these models are used*," while another indicated the value of the model, yet warned that the analysis "*becomes complicated including multiple variables.*" Hence, a multicriteria framework can facilitate the exploration of different combinations of solutions around which a consensus could coalesce. This points to the value in transparently discussing various trade-offs using user-centred approaches.

Policy and Politics of Energy Efficiency
Regarding policy solutions to support the implementation of the multi-criteria framework, the majority of stakeholders mentioned the need for CO_2 pricing, energy efficiency auctions, green loans and investments in new technologies, tax-free maintenance funds, and subsidies for various programmes, including energy-savings assessments, renovation, and insulation. In addition, stakeholders emphasised the need for energy efficiency standards, and enforcement and monitoring mechanisms for compliance with targets via regular energy reviews. Some argued that new measures are needed that better reflect environmental impact, such as CO_2 emissions per capita or energy consumption per capita, instead of energy consumption per square metre. Indeed, one participant mentioned incorporating "*other environmental impacts*" besides emissions along with "*social aspects, cultural aspects.*" Other solutions mentioned include improving the use of buildings, investing in skills and knowledge development for renovating and maintaining buildings, and improving public knowledge on savings, costs and emissions. To improve

the usability of the multicriteria model and implementation of its solutions, stakeholders emphasised that policymakers must support the use of such models via regulation and financial incentives. This points to the existence of opportunities for consensus building and mindset shifts about the wider benefits of energy efficiency measures by emphasising their social considerations.

10.3 Achieving Our Recommendation

The results from the participatory workshop show that multicriteria models can be useful framing tools to involve users in decisions about adopting efficiency measure. This is captured in our policy recommendation outlined in our title—*understand stakeholder perceptions and implementation possibilities for energy efficiency measures and policy through multicriteria modelling.* By highlighting both costs and benefits of efficiency measures, participatory workshops can facilitate dialogues among actors with diverse perspectives and help highlight feasible options for energy efficiency. Such discussions also create opportunities to envisage other aspects of the European Green Deal, such as improvements to social or health outcomes, for instance through the New European Bauhaus (European Commission, 2021).

Specifically, we recommend the following actions. *Decision-makers should present trade-offs, such as cost and emissions, and the combinations of acceptable solutions to various stakeholders such as housing associations, regulatory agencies, and financial institutions.* For instance, this approach could be integrated in the revised Energy Performance of Buildings Directive's (European Union, 2024) one-stop shops for the energy renovations of buildings to encourage stakeholder participation. Our analysis of trade-offs between costs and reductions in GHG emissions associated with energy efficiency measures shows that the choices available to stakeholders are much wider than generally perceived, and that investments in certain energy efficiency measures are able to reduce GHG emissions at a much higher rate than the associated cost increases. Therefore, stakeholders should be aware that certain measures can be both cost and emissions efficient, e.g. heat storage. Intermediate solutions also include wider benefits beyond the climate imperative, such as greater comfort for tenants and cost savings, for example, by being less dependent on market prices, as well as wider societal benefits, such as improved air quality and reduced stress on energy systems.

Furthermore, *decision-makers should adopt a user-centred approach to energy efficiency measures by encouraging stakeholder dialogues around decision-support tools to improve understanding of costs and benefits of various measures.* Recognising the difficulty of user engagement with policymaking, we recommend further embedding stakeholder dialogues into the formal political process at the stages of agenda setting, policy formulation, and implementation. Furthermore, structuring the stakeholder dialogue around the Pareto frontier proved to be fruitful. The multicriteria model served as a compelling visualisation and scenario tool, operationalising abstract ideas into a concrete set of plausible outcomes. Besides policymakers and housing associations, other organisations, e.g. regulatory authorities and financial institutions, that incorporate the perspectives of diverse stakeholders and conflicting objectives would benefit from taking a multicriteria perspective. Such consensus building has been amenable in real-world contexts, e.g. Coventry City Council (Montibeller & Franco, 2011).

Finally, *decision-makers should identify opportunities for consensus building and mindset shifts about the wider benefits of energy efficiency measures by emphasising their social considerations.* While there is a clear climate imperative for undertaking energy efficiency measures, stakeholders may be more motivated by other considerations. These could incorporate diversity, equity, and resilience. Although costs and emissions are also social considerations that receive the most attention, others would resonate with the public (Ewald et al., 2022) and should be addressed in such multicriteria assessments. Highlighting wider benefits of energy efficiency measures in consultations with stakeholders could be one way of identifying opportunities for consensus building around feasible options. Taking social considerations into account is particularly important for gaining acceptability for energy efficiency solutions.

Acknowledgements We are grateful to the feedback from attendees of a participatory workshop (11 December 2023).[4] This research was also funded by the Swedish Energy Agency, grant number P2022-01086.

---

[4] The agenda is available at https://stockholmuniversity.box.com/s/tpdebucmj11pk6y8b21uiq8nwnr60d4b

References

Carlander, J., & Thollander, P. (2023). Barriers to implementation of energy-efficient technologies in building construction projects—Results from a Swedish case study. *Resources, Environment and Sustainability, 11*, Article 100097. https://doi.org/10.1016/j.resenv.2022.100078

European Commission. (2019). *The European Green Deal. COM(2019) 640 final*. Brussels: European Commission.

European Commission. (2021). *New European Bauhaus. COMS(2021) 573 final*. Brussels: European Commission.

European Union. (2023). *Directive (EU) 2023/1791 of the European Parliament and of the Council of 13 September 2023 on energy efficiency and amending regulation (EU) 2023/955 (recast)*. Brussels: Official Journal of the European Union.

European Union. (2024). *Directive (EU) 2024/1275 of the European Parliament and of the Council of 24 April 2024 on the energy performance of buildings (recast)*. Brussels: Official Journal of the European Union.

Ewald, J., Sterner, T., & Sterner, E. (2022). Understanding the resistance to carbon taxes: Drivers and barriers among the general public and fuel-tax protesters. *Resource and Energy Economics, 70*, Article 101331. https://doi.org/10.1016/j.reseneeco.2022.101331

Huber, R. A., Maltby, T., Szulecki, K., & Ćetković, S. (2021). Is populism a challenge to European energy and climate policy? Empirical evidence across varieties of populism. *Journal of European Public Policy, 28*(7), 998–1017. https://doi.org/10.1080/13501763.2021.1918214

Mjörnell, K., Femenías, P., & Annadotter, K. (2019). Renovation strategies for multi-residential buildings from the record years in Sweden—Profit-driven or socioeconomically responsible? *Sustainability, 11*(24), Article 6988. https://doi.org/10.3390/su11246988

Montibeller, G., & Franco, L. A. (2011). Resource allocation in local government with facilitated portfolio decision analysis. In A. Salo, J. Keisler, & A. Morton (Eds.), *Portfolio decision analysis: Improved methods for resource allocation* (pp. 259–281). Springer. https://doi.org/10.1007/978-1-4419-9943-6_12

Phillips, L. D. (2011). The Royal Navy's Type 45 story: A case study. In A. Salo, J. Keisler, & A. Morton (Eds.), *Portfolio decision analysis: Improved methods for resource allocation* (pp. 53–75). Springer. https://doi.org/10.1007/978-1-4419-9943-6_3

Pineau, P. O. (2023). *L'équilibre énergétique*. Éditions Robert Laffont. ISBN: 9782924910306

Seibicke, H. (2024). Investigating stakeholder rationales for participating in collaborative interactions at the policy–science nexus. *Policy & Politics*. https://doi.org/10.1332/03055736Y2023D000000010

Siddiqui, A. S. (2024). *A multicriteria approach for a building energy management system*. Technical report. https://stockholmuniversity.box.com/s/9nesm21wbq9gmbwv2uuz30jil5nupooy

Von Platten, J., Sandels, C., Jörgensson, K., Karlsson, V., Mangold, M., & Mjörnell, K. (2020). Using machine learning to enrich building databases—Methods for tailored energy retrofits. *Energies*, *13*(10), Article 2574. https://doi.org/10.3390/en13102574

Winston, W. L., & Albright, S. C. (2018). *Practical management science* (6th ed.). Cengage.

Yeatts, D. E., Auden, D., Cooksey, C., & Chen, C. F. (2017). A systematic review of strategies for overcoming the barriers to energy-efficient technologies in buildings. *Energy Research & Social Science*, *32*, 76–85.

Open Access This chapter is licensed under the terms of the Creative Commons Attribution 4.0 International License (http://creativecommons.org/licenses/by/4.0/), which permits use, sharing, adaptation, distribution and reproduction in any medium or format, as long as you give appropriate credit to the original author(s) and the source, provide a link to the Creative Commons license and indicate if changes were made.

The images or other third party material in this chapter are included in the chapter's Creative Commons license, unless indicated otherwise in a credit line to the material. If material is not included in the chapter's Creative Commons license and your intended use is not permitted by statutory regulation or exceeds the permitted use, you will need to obtain permission directly from the copyright holder.

CHAPTER 11

Rethink Energy System Models to Support Interdisciplinary and Inclusive Just Transition Debates

Diana Süsser, Connor McGookin, Will McDowall, Francesco Lombardi, Lukas Braunreiter, and Stefan Bouzarovski

Policy Highlights To achieve the recommendation stated in the chapter title, we propose the following:

- Policymakers should demand more open and inclusive energy modelling processes to ensure that stakeholders can meaningfully contribute to the process.
- Policymakers should recognise the critical role of the Social Sciences and Humanities (SSH) in complementing energy modelling to receive a more holistic viewpoint on just pathways to climate neutrality. Both Science, Technology, Engineering and Mathematics

D. Süsser (✉) · S. Bouzarovski
Institute for European Energy and Climate Policy (IEECP), Amsterdam, The Netherlands
e-mail: diana@ieecp.org

© The Author(s) 2024
A. Crowther et al. (eds.), *Strengthening European Energy Policy*,
https://doi.org/10.1007/978-3-031-66481-6_11

(STEM) and SSH research is needed to transform our energy system to a just, climate-neutral future.
- Policymakers should establish cross- and transdisciplinary debates for incorporating more diverse voices into energy modelling.

Keywords Participatory modelling · Climate change mitigation · Energy justice · Just transition · Energy policy

11.1 Introduction

The goal of a just transition to climate neutrality is high on the political agenda. Just transition refers to "a fair and equitable process of transition to a post-carbon society" (McCauley & Heffron, 2018, p. 2). The concept has been recognised in the IPCC's latest mitigation assessment

---

S. Bouzarovski
e-mail: stefan.bouzarovski@manchester.ac.uk

C. McGookin
School of Sustainable Energy Engineering, Simon Fraser University, Vancouver, BC, Canada
e-mail: connor_mcgookin@sfu.ca

W. McDowall
UCL Institute for Sustainable Resources, University College London, London, UK
e-mail: w.mcdowall@ucl.ac.uk

F. Lombardi
Faculty of Technology, Policy and Management, TU Delft, Delft, The Netherlands
e-mail: f.lombardi@tudelft.nl

L. Braunreiter
Swiss Energy Foundation, Zürich, Switzerland

S. Bouzarovski
Department of Geography, The University of Manchester, Manchester, UK

and the European Green Deal. The European Union has set stringent net-zero greenhouse gas emission targets, while also declaring it will leave no person and no place behind.

The dominant tools for understanding the energy transition are energy system models (Süsser et al., 2021a). The most prominent whole system approaches—Energy System Optimisation Models (ESOMs) and Integrated Assessment Models (IAMs)—are important policy tools. These provide a representation of current and future emissions across different scales; pan-EU, national, and regions or cities. While they have great value in providing techno-economic least-cost pathways for decarbonisation, we argue that their ability to reflect the real-world energy transition is limited. Two critical gaps we see are that, firstly, the models are not designed to reflect important aspects of fairness and inclusion, and secondly, they tend to assume very little or no changes to social and political institutions (e.g. future energy demand is generally based on projecting continuous GDP growth).

Consequently, current modelling practices are often incompatible with the goal of a just transition. The models, optimising for least cost, are unlikely to produce equitable outcomes, and modelling teams have tended not to focus on equity or just transition issues (Sonja & Harald, 2018). Recently, there has been some interest in ways to incorporate broader societal considerations into modelling tools (Krumm et al., 2022; Lonergan et al., 2023). This includes using existing models to scrutinise narratives, intensifying collaboration across scholars, or structurally modifying and building new models to integrate Social Science research (Holtz et al., 2015; Trutnevyte et al., 2019). Nevertheless, more effort is needed to increase inclusive participation in modelling processes and to integrate aspects of fairness or justice in energy modelling (Lonergan et al., 2023; McGookin et al., 2021).

The arguments made in this chapter stem from the context of three EU-funded research projects, SENTINEL, SEEDS and JustWind4All, as well as on an online discussion between the author team (September 2023) (McGookin et al., 2024a), a joint workshop with the International Renewable Energy Agency (IRENA) on stakeholder-driven scenario development for just transitions to climate neutrality (November 2023) (Süsser, 2024; Süsser & Goussous, 2024), and feedback from a presentation at the Behave 2023 conference (November 2023). At the workshops and discussions, both Social Scientists and Computer Engineers were present, who work in research, practice, and policy, including governmental authorities, international development agencies and the energy industry. We discussed current gaps in modelling practices and solutions to improve tools and modelling processes. By building on these combined insights, we argue that integrated and complementary energy modelling and Social Sciences research are crucial to enable equitable pathways to climate neutrality. Policy would benefit from insights based not only on modelled techno-economic pathways, but also on the results of debates with the stakeholders[1] and citizens. To achieve just transitions, models must be complemented with Social Science research, including Policy Research, Psychology and Human Geography, to open up debates and enable better informed decision-making.

11.2 Insights on Modelling Gaps and Ways to Improve and Complement Energy Modelling

In this chapter, we apply a justice lens to energy modelling (Table 11.1). This is done using three energy justice principles: *distributional justice*, which focuses on the equitable distribution of costs and benefits; *procedural justice*, which refers to transparent decision-making processes and adequate representation; and *recognitional justice*, which acknowledges past injustices and ongoing risks of underrepresentation (Jenkins et al., 2016; Walker & Day, 2012).

11.2.1 What Are the Limitations of Models and Modelling Approaches?

A key limitation of current energy models is their grounding in techno-economic worldviews that prioritise total costs rather than distributions, and which obscure procedural and recognitional dimensions of justice. Models are navigated through modellers' frameworks, norms, and values, which often remain inherently ambiguous (Silvast et al., 2020). A narrow techno-economic lens pushes into the background alternative perspectives that might challenge foundational assumptions. Models are shaped by

---

[1] We define stakeholders as all those affected by or interested in the energy transition, including policymakers, the energy industry and civil society organisations.

Table 11.1 How Social Sciences can contribute to filling modelling gaps on energy justice

Justice dimensions	Gaps in modelling	Contributions of Social Sciences to close gaps
Recognition: Whose worldviews are represented and excluded?	The techno-economic perspective in models limits potential for recognition of diverse groups	Challenges worldviews and link them to policy goals; opens-up processes to diverse perspectives
Procedure: Who is involved in the modelling process? What models are used? How is modelling used to inform decisions?	Models can reduce the space for debate and dissent, excluding marginalised voices and 'de-politicising' debate. The process of model development is rarely transparent. Public rarely participates in modelling processes	Challenges the assumptions behind models and what is missing from them; raises new questions; provides participatory research expertise; communicates model uncertainties and outcomes
Distribution: How are distributional impacts assessed? Who will be the (local/regional) winners and losers of the transition?	Models typically explore distributional consequences as second-order concern, if at all	Investigates local and regional transition impacts; documents people's lived experience of the energy transition; includes human behaviour and responses

certain societal discourses, which are reproduced and reinforced (Ellenbeck & Lilliestam, 2019). As such, models may become engines of injustice and exclusion themselves.

Models can 'depoliticise' debates, undermining procedural justice. They do this first by narrowing the frame of debate, as they provide only a simplified representation of reality. In doing so, they push excluded perspectives into the background, privileging some issues and perspectives over others. There is a basic trade-off here: such narrowing is important for tractability and 'closure' around a particular problem framing, but this comes at the cost of respect for plural perspectives (Stirling, 2008). Second, models often have power in debate (Aykut et al., 2019). Their purported accuracy, technical complexity, and association with 'objective science' lend them strong credibility (Porter, 2020), even when it is not clear what the knowledge claims arising from a given model might be. The risk is that the space for political dialogue is removed: the apparently objective model, which only few are competent to critique, both frames the debate in ways that exclude certain perspectives and obscures many of the normative and political judgements that underpin the conclusions.

Model development processes are not transparent and are rarely informed by co-design or participation, further limiting procedural fairness. Despite calls for the opening up of energy system models (Morrison, 2018; Pfenninger, 2017), the assumptions that determine modelling outputs remain opaque. Progress with open modelling has reduced this concern in recent years, but it remains true that relatively few people have the skills required to unravel the assumptions underpinning certain findings and be able to challenge them. This leaves a significant amount of control over the framing and the logic within the modelling team. Moreover, modelling processes are rarely opened-up to wider participatory and co-design processes (McGookin et al., 2021). We argue that only if different stakeholders are part of the modelling process, they can influence it, and thus, ways to build-in stakeholder perspectives can be explored.

Models overlook transition impacts. Some modelling studies do account for distributional impacts, such as which regions stand to benefit or lose from the transition (Caulfield et al., 2022; Li et al., 2016; McDowall et al., 2023). However, they typically focus on economic vulnerabilities and examine distributional issues as a consequence of least-cost pathways, implying that distributional issues are secondary in importance to total costs (for a rare exception, see Sasse & Trutnevyte, 2020). They contribute to a frame in which difficult distributional impacts are seen as the unfortunate, but necessary, consequence of the least-cost transition path, rather than opening a conversation about society's prioritisation of inequitable outcomes.

11.2.2 How Can Social Sciences Address Modelling Gaps?

Recognitional Justice
Social Sciences can challenge dominant worldviews by discussing mental models behind the computer-based models. Models are built from assemblages of theory, data and (often tacit) social norms about how the world 'works'. Insights from Behavioural Science, Political Science and other fields can unpack those assumptions, and thus open up the possibility for model-based explorations of more radical or emancipatory futures. A recognitional justice lens demands that analyses recognise diverse perspectives, values, and aspirations by "engag[ing] with other knowledge systems as active contributors of solutions" (Rubiano Rivadeneira &

Carton, 2022, p. 8). For example, visioning documents developed at the community-level have been shown to provide context-based nuance that challenges the techno-managerial "indexification of poverty" (Kiely & Strong, 2023, p. 1758)—i.e. the use of statistical indices to measure poverty—and provide alternative ways of building energy poverty models. Such approaches can contribute to public debates on possible and desirable energy futures, and systematically rebalance existing power relations within the energy system, promoting 'recognitional' and 'procedural' justice. This can be crucial not only to improve public participation in climate and energy policymaking, but also to increase trust in the policy outcomes.

Procedural Justice
Social Scientists can help challenge the assumptions behind models. To enable cross-disciplinary dialogues, modelling processes must be transparent. This is not simply a matter of open code, open data, and good documentation—important though these are. Modelling data and assumptions should be discussed within interdisciplinary teams, and also with stakeholders, to create a better understanding of the importance of assumptions and uncertainties in modelling (McGookin et al., 2024b). Transparency must be an ongoing process that ensures models are continually being explained, challenged and critiqued. Social Sciences can thus help to redress the power imbalances created by complex modelling tools.

Modelling perspectives can be expanded with Social Science research to better understand social aspects, such as attitudes towards different energy futures or lived energy experiences. This was attempted, for instance, in the SEEDS project, where stakeholder needs were used to expand the default outputs provided by models, to better reflect stakeholder concerns. Furthermore, Behavioural Science and Psychology can provide theories and evidence on behavioural change or people's preferences, which can be used in modelling tools. An example is provided by the SENTINEL project, where social-political storylines based on different governance logics and social and political observations (Süsser et al., 2021b) constrained feasible net-zero configurations of the European energy system (Mayer et al., 2024). Using models alongside other processes can ensure that broader perspectives are included in the analysis (McDowall, 2014).

Social Scientists can provide participatory research and communication expertise to modellers. Instead of modellers re-inventing the wheel, they should seek to work with these experts through transdisciplinary approaches. Visualisations of modelling results can facilitate policy dialogue, and the communication of model uncertainties and assumptions is critical to create an understanding among modelling users, including policymakers, what model outcomes mean and what they do not mean. Moreover, Social Scientists can provide insights into participatory methods and how to plan effective public engagement processes as an integral part of modelling. For example, the community engagement in the modelling work by McGookin et al. (2022) benefited greatly from Social Science perspectives. The research team implemented a broader engagement process to explore what a sustainable future for the area would look like, resulting in several important local projects.

Distributional Justice
Social Science can contribute to a better understanding of how positive and negative impacts of the transition are distributed. Regions and communities will be affected differently by the transition, depending on their social and geographic circumstances, the current status of the transition, and capacities to respond, among others. This requires models to account for existing regional differences and potential underlying injustices in the energy transformation. For example, a modelling study by Mayer et al. (2024) showed that positive employment effects could lead to higher welfare levels, which would otherwise have been neglected if only the costs of energy system configurations had been considered. Local and regional analyses could be used to assess the impact of transitions, including costs and benefits, and provide important insights to complement modelling tools.

11.3 Achieving Our Recommendation

As per the title of this chapter, our core recommendation is that policy should: *rethink energy system models to support interdisciplinary and inclusive just transition debates*. This recommendation is underpinned by three sub-recommendations:

First, *policymakers should demand more open and inclusive energy systems modelling processes*. Diverse perspectives can contribute to a critical reflection of current injustices in the energy transition and their anchoring in models. Addressing existing injustices and ensuring fairness and inclusiveness in the energy transformation is critical to achieve the energy policy goals for a just transition to climate neutrality. Thus, policymaking should require open, transparent, participatory modelling processes from the modelling community and work with institutes that align with this standard. Such processes should facilitate a critical engagement with and around modelling tools, as well as building a better understanding of the 'power' of model assumptions and model limitations. Policymakers should initiate and/or fund research programmes that require the formation of interdisciplinary research teams with diverse expertise, the convening of participatory modelling processes, or stakeholder-based committees or partnerships.

Second, *policymakers should recognise the critical role of the Social Sciences in complementing energy systems modelling to receive a more holistic viewpoint on just pathways to climate neutrality*. A constructive critique of models and modelling processes is required, which may highlight injustices or lack of attention to justice issues. This requires the EU funding of research and practice projects that produce critical socio-psychological and institutional insights, such as how to meaningfully engage the public in energy infrastructure projects, or perceptions and needs for transitions away from coal and carbon-intensive industries. This would contribute to the achievement of policy goals to accelerate the expansion of renewable energy, in line with the 'Fit for 55'-package, and to support regions that are most vulnerable to the transition under the Just Transition Mechanism.

Third, *policymakers should establish cross- and transdisciplinary debates for incorporating more diverse voices into energy systems modelling*. There is not only one energy future; visions, values, and aspirations of researchers with different backgrounds, as well as those from diverse stakeholders and citizens, can inform the development of alternative storylines and scenarios. McGookin et al. (2024b) have suggested best practice guidelines for incorporating diverse voices into energy modelling. However, modelling projects are often restricted by funders' requirements, which may prevent engagement in deliberative activities. Policymakers—and in particular funders of modelling—should create spaces for cross-disciplinary and participatory dialogue to open up modelling. In deliberative dialogues, models can function as 'exploration tools'—helping to foster debate, rather than replace it.

Acknowledgement Stefan Bouzarovski gratefully acknowledges funding from the Energy Demand Research Centre (EDRC), supported by the Engineering and Physical Sciences Research Council and the Economic and Social Research Council [grant number EP/Y010078/1].

References

Aykut, S., Demortain, D., & Benboudiz, B. (2019). The politics of anticipatory expertise: Plurality and contestation of futures knowledge in governance— Introduction to the special issue. *Science & Technology Studies, 32*(4), 2–12. https://doi.org/10.23987/sts.87369

Caulfield, B., Furszyfer, D., Stefaniec, A., & Foley, A. (2022). Measuring the equity impacts of government subsidies for electric vehicles. *Energy, 248*, 123588. https://doi.org/10.1016/j.energy.2022.123588

Ellenbeck, S., & Lilliestam, J. (2019). How modelers construct energy costs: Discursive elements in Energy System and integrated assessment models. *Energy Research & Social Science, 47*, 69–77. https://doi.org/10.1016/j.erss.2018.08.021

Holtz, G., Alkemade, F., De Haan, F., Köhler, J., Trutnevyte, E., Luthe, T., Halbe, J., Papachristos, G., Chappin, E., Kwakkel, J., & Ruutu, S. (2015). Prospects of modelling societal transitions: Position paper of an emerging community. *Environmental Innovation and Societal Transitions, 17*, 41–58. https://doi.org/10.1016/j.eist.2015.05.006

Jenkins, K., McCauley, D., Heffron, R., Stephan, H., & Rehner, R. (2016). Energy justice: A conceptual review. *Energy Research & Social Science, 11*, 174–182. https://doi.org/10.1016/j.erss.2015.10.004

Kiely, E., & Strong, S. (2023). The Indexification of poverty: The covert politics of SMALL-AREA indices. *Antipode, 55*(6), 1758–1780. https://doi.org/10.1111/anti.12959

Krumm, A., Süsser, D., & Blechinger, P. (2022). Modelling social aspects of the energy transition: What is the current representation of social factors in energy models? *Energy, 239*, 121706. https://doi.org/10.1016/j.energy.2021.121706

Li, F. G. N., Pye, S., & Strachan, N. (2016). Regional winners and losers in future UK energy system transitions. *Energy Strategy Reviews, 13–14*, 11–31. https://doi.org/10.1016/j.esr.2016.08.002

Lonergan, K. E., Suter, N., & Sansavini, G. (2023). Energy systems modelling for just transitions. *Energy Policy, 183*, 113791. https://doi.org/10.1016/j.enpol.2023.113791

Mayer, J., Süsser, D., Pickering, B., Bachner, G., & Sanvito, F. D. (2024). Economy-wide impacts of socio-politically driven net-zero energy systems in Europe. *Energy, 291*, 130425. https://doi.org/10.1016/j.energy.2024.130425

McCauley, D., & Heffron, R. (2018). Just transition: Integrating climate, energy and environmental justice. *Energy Policy, 119*, 1–7. https://doi.org/10.1016/j.enpol.2018.04.014

McDowall, W. (2014). Exploring possible transition pathways for hydrogen energy: A hybrid approach using socio-technical scenarios and energy system modelling. *Futures, 63*, 1–14. https://doi.org/10.1016/j.futures.2014.07.004

McDowall, W., Reinauer, T., Fragkos, P., Miedzinski, M., & Cronin, J. (2023). Mapping regional vulnerability in Europe's energy transition: Development and application of an indicator to assess declining employment in four carbon-intensive industries. *Climatic Change, 176*(2), 7. https://doi.org/10.1007/s10584-022-03478-w

McGookin, C., Mac Uidhir, T., Gallachóir, Ó., & B., & Byrne, E. (2022). Doing things differently: Bridging community concerns and energy system modelling with a transdisciplinary approach in rural Ireland. *Energy Research & Social Science, 89*, 102658. https://doi.org/10.1016/j.erss.2022.102658

McGookin, C., Mac Uidhir, T., Ó Gallachóir, B., & Byrne, E. (2021). Participatory methods in energy system modelling and planning—A review. *Renewable and Sustainable Energy Reviews, 151*, 111504. https://doi.org/10.1016/j.rser.2021.111504

McGookin, C., Bouzarovski, S., Braunreiter, L., Lombardi, F., McDowall, W., & Süsser, D. (2024a). *Results from the online workshop of the author team*. Zenodo.https://doi.org/10.5281/zenodo.11162349

McGookin, C., Süsser, D., Xexakis, G., Trutnevyte, E., McDowall, W., Nikas, A., Koasidis, K., Few, S., Andersen, P. D., & Demski, C. (2024b). Advancing participatory energy systems modelling. *Energy Strategy Reviews, 52*, 101319. https://doi.org/10.1016/j.esr.2024.101319

Morrison, R. (2018). Energy system modeling: Public transparency, scientific reproducibility, and open development. *Energy Strategy Reviews, 20*, 49–63. https://doi.org/10.1016/j.esr.2017.12.010

Pfenninger, S. (2017). Energy scientists must show their workings. *Nature, 542*(7642), 393–393. https://doi.org/10.1038/542393a

Porter, T. M. (2020). Trust in numbers: The pursuit of objectivity in science and public life. *Princeton University Press*. https://doi.org/10.2307/j.ctvxcrz2b

Rubiano Rivadeneira, N., & Carton, W. (2022). (In)justice in modelled climate futures: A review of integrated assessment modelling critiques through a justice lens. *Energy Research & Social Science, 92*, 102781. https://doi.org/10.1016/j.erss.2022.102781

Sasse, J.-P., & Trutnevyte, E. (2020). Regional impacts of electricity system transition in Central Europe until 2035. *Nature Communications*, *11*(1), 4972. https://doi.org/10.1038/s41467-020-18812-y

Silvast, A., Laes, E., Abram, S., & Bombaerts, G. (2020). What do energy modellers know? An ethnography of epistemic values and knowledge models. *Energy Research & Social Science*, *66*, 101495. https://doi.org/10.1016/j.erss.2020.101495

Sonja, K., & Harald, W. (2018). Building equity in: Strategies for integrating equity into modelling for a 1.5 °C world. *Philosophical Transactions of the Royal Society A: Mathematical, Physical and Engineering Sciences*, *376*(2119), 20160461. https://doi.org/10.1098/rsta.2016.0461

Stirling, A. (2008). "Opening Up" and "Closing down": Power, participation, and pluralism in the social appraisal of technology. *Science, Technology, & Human Values*, *33*(2), 262–294. https://doi.org/10.1177/0162243907311265

Süsser, D. (2024). *Results from the joint IRENA and IEECP workshop: Stakeholder-driven scenarios for a just transition to climate neutrality*. Zenodo.https://doi.org/10.5281/zenodo.11162399

Süsser, D., Ceglarz, A., Gaschnig, H., Stavrakas, V., Flamos, A., Giannakidis, G., & Lilliestam, J. (2021a). Model-based policymaking or policy-based modelling? How energy models and energy policy interact. *Energy Research & Social Science*, *75*, 101984. https://doi.org/10.1016/j.erss.2021.101984

Süsser, D., al Rakouki, H., & Lilliestam, J. (2021b). *The QTDIAN modelling toolbox–Quantification of social drivers and constraints of the diffusion of energy technologies. Deliverable 2.3. Version 1. Sustainable Energy Transitions Laboratory (SENTINEL) project*. Potsdam: Institute for Advanced Sustainability Studies (IASS). https://doi.org/10.48481/iass.2021.015

Süsser, D., & Goussous, N. (2024). *IRENA and IEECP Joint Workshop: Stakeholder-driven energy scenarios for a just transition: Dialogue with the scientific community*. Zenodo.https://doi.org/10.5281/zenodo.11162529

Trutnevyte, E., Hirt, L. F., Bauer, N., Cherp, A., Hawkes, A., Edelenbosch, O. Y., Pedde, S., & Van Vuuren, D. P. (2019). Societal transformations in models for energy and climate policy: The ambitious next step. *One Earth*, *1*(4), 423–433. https://doi.org/10.1016/j.oneear.2019.12.002

Walker, G., & Day, R. (2012). Fuel poverty as injustice: Integrating distribution, recognition and procedure in the struggle for affordable warmth. *Energy Policy*, *49*, 69–75. https://doi.org/10.1016/j.enpol.2012.01.044

Open Access This chapter is licensed under the terms of the Creative Commons Attribution 4.0 International License (http://creativecommons.org/licenses/by/4.0/), which permits use, sharing, adaptation, distribution and reproduction in any medium or format, as long as you give appropriate credit to the original author(s) and the source, provide a link to the Creative Commons license and indicate if changes were made.

The images or other third party material in this chapter are included in the chapter's Creative Commons license, unless indicated otherwise in a credit line to the material. If material is not included in the chapter's Creative Commons license and your intended use is not permitted by statutory regulation or exceeds the permitted use, you will need to obtain permission directly from the copyright holder.

PART VI

Conclusion

CHAPTER 12

Reflections on Interdisciplinary Collaborations for European Energy Policy and Governance

Ami Crowther, Chris Foulds, Rosie Robison, and Ganna Gladkykh

Abstract The policy recommendations presented in this book demonstrate the value and opportunities of interdisciplinarity for policymaking. The recommendations produced cover a diverse range of topics and policy areas and are informed by various interdisciplinary collaborative activities. The outputs of the interdisciplinary collaborations evidence: (1) how questions related to energy supply, demand and systems benefit

---

A. Crowther (✉) · C. Foulds · R. Robison
Global Sustainability Institute, Anglia Ruskin University, Cambridge, UK
e-mail: ami.crowther@aru.ac.uk

C. Foulds
e-mail: chris.foulds@aru.ac.uk

R. Robison
e-mail: rosie.robison@aru.ac.uk

G. Gladkykh
European Energy Research Alliance, Brussels, Belgium
e-mail: ganna.gladkykh@sei.org

© The Author(s) 2024
A. Crowther et al. (eds.), *Strengthening European Energy Policy*,
https://doi.org/10.1007/978-3-031-66481-6_12

from both Social Sciences and Humanities (SSH) and Science, Technology, Engineering and Mathematics (STEM) perspectives), (2) that the achievement of EU policies can require the participation of multiple actors across multiple scales, (3) how policymaking can be supported not only through research on policy topics, but also by research on the processes of policymaking and their governance environments, and (4) the complex negotiation processes that exist at research-policy interfaces. Yet, there is the need to consider interdisciplinarity, and what this means in practice, critically. In doing so, it demonstrates the value of focusing more on the interdisciplinary processes and experiences in play. Indeed, across the book chapters, there were commonalities in how the interdisciplinary collaborations occurred in practice. This book not only demonstrates the value and opportunities for interdisciplinary research, but also, we hope, will encourage others to engage in interdisciplinary activities.

Keywords Policy · Governance · Interdisciplinary · SSH-STEM collaboration

12.1 Reflections on Policy and Governance Recommendations

The purpose of this book was to strengthen European energy policy by generating concrete interdisciplinary recommendations for relevant EU energy strategies. As such, the titles of the core chapters (Chapters 2–11) are the policy recommendations generated, with this clearly communicating the key messages being discussed within these chapters. An overview of said recommendations, and how they align (even if lightly at times) to current European Union (EU) and European Commission (EC) agendas, is provided in Table 12.1. The chapters cover a range of policy areas and topics, including retrofit, energy communities, and digital infrastructures, and draw upon various methods to explore the topics covered, including modelling, workshops, and literature reviews. The outputs of these chapters demonstrate the value of bringing together researchers from the Social Sciences and Humanities (SSH) and Science, Technology, Engineering and Mathematics (STEM), with this being captured in our reflections on the chapter recommendations.

Table 12.1 Chapter policy recommendations and related EU policy instrument(s)

Cluster	Chapter number	Recommendation/title	EU policy instrument(s)
Part I. Navigating community participation	2	Simplify the uptake of community energy by leveraging intermediaries and the use of digital planning tools	• Clean energy for all Europeans • Renewable energy directive
	3	Prioritise inclusive, early and continuous societal engagement to maximise the benefits of geothermal technologies	• REPowerEU • Renewable Energy Directive
	4	Create a co-learning environment for Geothermal Energy Communities across the European and African Unions	• Clean energy for all Europeans • Renewable Energy Directive • Africa-EU Energy Partnership
Part II. Navigating knowledges for the built environment	5	Facilitate the development of energy literacy amongst citizens to support their meaningful participation in the energy transition	• REPowerEU • Energy Performance of Buildings Directive • National Building Renovation Plans
	6	Support place-based and inclusive supply chain, employment and skills strategies for housing-energy retrofit	• A renovation wave for Europe • Energy Performance of Buildings Directive • EU Energy Efficiency Directive

(continued)

Table 12.1 (continued)

Cluster	Chapter number	Recommendation/title	EU policy instrument(s)
Part III. Navigating the delivery of new technologies	7	Promote integrated policy design to overcome social and technical challenges for agrivoltaic deployment	• REPowerEU
	8	Increase social acceptability of nuclear fusion, agrivoltaics and offshore wind through national support programmes	• REPowerEU • Net-Zero Industry Act (NZIA)
	9	Protect the EU's digital energy infrastructure against cyberthreats through advanced technologies, human vulnerability mitigation and ethical practices	• REPowerEU • EU Digital Single Market Strategy • Critical Infrastructure Directive • Cybersecurity Act • Network and Information Security (NIS) Directive
Part IV. Navigating models for policy development	10	Understand stakeholder perceptions and implementation possibilities for energy-efficiency measures and policy through multicriteria modelling	• EU Cities Mission • Energy Performance of Buildings Directive • Energy Efficiency Directive • National Building Renovation Plans
	11	Rethink energy system models to support interdisciplinary and inclusive just transition debates	• EU Just Transition Mechanism

From our own experience, we have found that a common misconception is that SSH research is better equipped to research energy demand, whilst STEM approaches align more with questions of energy supply (in spite of SSH literatures dealing with systems of provision, and similarly STEM literatures dealing with demand technologies). This book provides evidence that this misconception is not true and, as such, interdisciplinarity across a range of both supply and demand policy areas can be very productive.

Although the starting point for the interdisciplinary chapters was to develop policy recommendations for the EU, the implementation of many of the recommendations presented requires the involvement of different actors at different scales. This reflects the different foci of the chapters, looking at both energy demand and supply, as well as some system-level topics. Across the chapters reference is made to the role of Member States, energy sector organisations, individual households, amongst others, in achieving the policy recommendations. As such, some of the EU policy recommendations presented require coordination across multiple scales and governance structures.

Many chapters focus on policymaking processes and the structural landscapes in which the policy is eventually to be implemented. Chapters also considered how wider governance initiatives may complement existing policy initiatives. As such, there was implicit interest in going beyond policy targets or intended outcomes, into, for example, how policymaking gets done, by whom, and within what real-world contexts. This also connects to the view that policymaking is a process co-owned by multiple actors, requiring policy-supporting tools and interaction between policymakers, researchers, professional communities, citizens and other stakeholders.

Furthermore, the chapters' recommendations are situated within a broader set of negotiation processes that exist at research-policy interfaces. Indeed, there is an ongoing balancing act, whereby on the one side: policymakers would ideally like concrete recommendations, which fit within their paradigm and thus their existing policy commitments and delivery mechanisms, and on the other side, researchers are keen to epistemically experiment and push the boundaries of what is possible (e.g. in terms of policy evidence) through innovative interdisciplinary collaborations, but are wary of their research being instrumentalised. The challenge then is how these agendas align.

12.2 Reflections on Innovation in Interdisciplinary Collaborations

It is clear that this book represents a significant interdisciplinary undertaking. Indeed, of the 57 authors contributing to the policy recommendation chapters (Chapters 2–11), 29 classified themselves as SSH researchers and 28 as STEM. SSH disciplines ranged, for example, from Environmental Social Science and Sociology to Marketing and Business Management; with STEM disciplines ranging from Computer Science and Physics to various Engineering and Geoscience disciplines.

However, disciplines are fluid and their boundaries are porous; they are not rigid institutional entities (Jacobs, 2013). Thus, it quickly became apparent to us that asking the chapter teams for disciplinary self-classifications was unlikely to have been straightforward for them. For example, we have observed how many authors: have individual track records straddling both STEM and SSH disciplines; are based in very applied policy or practice settings, where disciplinary identities matter much less; and/or, assign themselves to relatively new hybrid interdisciplines, such as Gender Studies or Urban Studies. In these ways, whilst disciplines are "useful proxies for different ways of generating, interpreting and applying knowledges, we should not obsess about" (Silvast & Foulds, 2022, p. 8) counting disciplines or attempting to draw precise disciplinary boundaries. Instead, we argue for more reflection upon the actual interdisciplinary processes and experiences in play.

The enactment of interdisciplinarity usually hinges on what is regarded to be epistemically palatable amongst the group of collaborators (Silvast & Foulds, 2022). It is interesting then to observe the positions that STEM and SSH perspectives took during the collaboration. Herein, STEM contributions would usually set the tone for what was technically possible in the future or what the current starting arrangements were. For example, Calver et al. (Chapter 5) and Macrorie et al. (Chapter 6) used STEM to describe technical understandings of the energy system and construction industry, respectively. Whereas, SSH contributions would often be more exploratory, using SSH researcher skills, such as problematisation, critique, empathy, reflexivity, etc., to orient interdisciplinary discussions towards societal needs. SSH would therefore commonly be used to open up the traditional STEM positions. For example, Rohse et al. (Chapter 3) used SSH insights to unpack different ways of thinking about engagement in geothermal projects.

Refreshingly, we are pleased to observe that, of our 10 chapters, seven were led by SSH researchers. This is a welcome change, given the traditional dominance of STEM in SSH-STEM interdisciplinarity (Kropp, 2021). We would speculate that this pushback against the normal scientific hierarchy of knowledge-making may be behind four trends we observed across the chapters: (1) STEM sometimes took a more subordinate role, (2) alternative offerings of SSH were made clearer (e.g. criticality, reflexivity), (3) we enjoyed some constructive resistance and critique from the teams towards the normative ambitions of the book project, and (4) the conclusions and recommendations were overall slightly more tentative (e.g. less definitive, replicable and more contextually-grounded).

12.3 Closing Remarks: Interdisciplinarity for Strengthening Energy Policy and Governance

The process of developing this book—from launching the initial call, to working with the interdisciplinary chapter teams, to engaging with Foreword and Afterword authors—has highlighted the growing interest from researchers, policymakers, and others, in collaborative SSH-STEM research. It has also shown the need for better mechanisms to facilitate the collaboration of SSH and STEM disciplines to evaluate, and ultimately strengthen, European energy policy and governance. The combination of different perspectives and methodologies across the chapters has supported critical engagement with current EU energy ambitions and policies, and has identified ways in which they can be enhanced, which could not be achieved if remaining in disciplinary silos.

Although the primary aim of this book project was to produce concrete EU energy policy recommendations through interdisciplinary collaborations, a secondary ambition was to showcase the value and opportunities of interdisciplinarity and inspire others to undertake their own interdisciplinary activities. Whilst we appreciate that this book does not capture the challenges, and potentially difficult conversations, of undertaking interdisciplinary work, it does demonstrate how the combination of different perspectives can develop insights that address energy challenges. As such, we encourage others to undertake more interdisciplinary activities—whether that be policymakers engaging with different communities and perspectives to inform their activities; SSH researchers reaching out to their STEM colleagues down the corridor, and vice versa; or reaching out to someone new to provide support as you expand the remit of your work. After all, by moving beyond disciplinary silos, engaging with different perspectives, and adopting different methods, it helps stimulate innovative approaches and alternative ways of thinking, which are critical for addressing the complex sustainability and energy transition challenges we face.

References

Jacobs, D. (2013). Policy invention as evolutionary tinkering and codification: The emergence of feed-in tariffs for renewable electricity. *Environmental Politics*, 23(5), 755–773. https://doi.org/10.1080/09644016.2014.923627

Kropp, K. (2021). The EU and the social sciences: A fragile relationship. *The Sociological Review*, 69(6), 1325–1341. https://doi.org/10.1177/003802 61211034706

Silvast, A., & Foulds, C. (2022). *Sociology of interdisciplinarity: Dynamics of energy research*. Palgrave Macmillan.

Open Access This chapter is licensed under the terms of the Creative Commons Attribution 4.0 International License (http://creativecommons.org/licenses/by/4.0/), which permits use, sharing, adaptation, distribution and reproduction in any medium or format, as long as you give appropriate credit to the original author(s) and the source, provide a link to the Creative Commons license and indicate if changes were made.

The images or other third party material in this chapter are included in the chapter's Creative Commons license, unless indicated otherwise in a credit line to the material. If material is not included in the chapter's Creative Commons license and your intended use is not permitted by statutory regulation or exceeds the permitted use, you will need to obtain permission directly from the copyright holder.

Afterword 1: A Quest for More Intentional Interdisciplinary Synergies by Giulia Sonetti and Osman Arrobbio

Giulia Sonetti is a Researcher and Project Manager at the Research Institute for Sustainability Science and Technology, Universitat Politècnica de Catalunya—Barcelona Tech. Her work focuses on transformative learning, transdisciplinarity, and sustainability education, with a special focus on mental health, climate anxiety, and active hope in future generations.

Osman Arrobbio is an Assistant Professor in Sociology of the Environment at the University of Parma (Italy), ESH Lab (Environmental Social Humanities Lab). He obtained his PhD in Sociology at the University of Turin. Among his main research interests are: energy transitions, ecological transitions, sufficiency, practice theory, light pollution.

In this edited book on EU energy policy aimed at transitioning to carbon neutrality, the convergence of Social Sciences and Humanities (SSH) with Science, Technology, Engineering and Mathematics (STEM) enriches the dialogue on sustainable energy strategies. The chapters, each through a distinct lens, cohesively advocate for interdisciplinary approaches essential for addressing complex environmental and societal challenges.

For instance, in Chapter 10 by Buylova et al., the authors explore energy efficiency within urban planning, employing multicriteria decision-making models that encapsulate both economic and environmental objectives. This approach exemplifies the necessity of frameworks that operationalise the trade-offs in energy policy decisions, in terms that resonate with multiple stakeholders including housing associations and regulatory bodies. Similarly, in Chapter 4, Büscher et al. discuss the potential of geothermal energy communities in both the European Union and African Union contexts, illustrating the global applicability of locally-rooted energy solutions.

A central theme across the book is the pivotal role of communities and the need for involving a broad array of stakeholders in the energy transition process (Arrobbio & Sonetti 2021). This reflects a broader objective mirrored in SSH CENTRE's mission to foster inclusive and practical cross-sectoral collaborations. Notably, in Chapter 6, Macrorie et al. underscore the importance of developing skilled and diverse workforces for building retrofit projects, which highlights another critical aspect of community involvement in sustainability transitions.

Indeed, the integration of SSH insights with STEM expertise within Chapter 6 significantly enhances the depth and applicability of the research presented. It integrates insights from Civil and Structural Engineering, Human Geography, and Sociology. This interdisciplinary perspective explores how different approaches can support inclusivity in skills provision and employment outcomes, and thereby fundamentally illustrates the potential fruits of interdisciplinarity.

To mainstream such interdisciplinary approaches, it is imperative that educational and funding structures within academic and policy-making realms support such collaborative ventures (Bina et al., 2021), acknowledging the nuances and ramifications of cross-disciplinary scientific work, from individual academics to broader systemic factors like funding, resource allocation, and power dynamics that influence interdisciplinary research—as per a Sociology of Interdisciplinarity (Silvast & Foulds, 2022). In this line, programmes that nurture a combination of SSH and STEM—and hopefully transdisciplinary competencies among upcoming professionals—will be essential in sustaining the innovative momentum required to tackle contemporary energy challenges.

Looking forward, the insights from this book should guide the next phases of research and action in EU energy policy. The strategies discussed, such as those in Chapter 8 by Clain et al. about social acceptability of energy transitions, need to be scaled and adapted to meet the diverse challenges across EU Member States. Policymakers, in collaboration with researchers and industry leaders, must take the lead in championing the implementation of these interdisciplinary strategies, ensuring they translate from theoretical frameworks (and relative EU call wordings and topic destinations) to the selection of funded solutions (Sonetti et al., 2020).

Moreover, ongoing dialogue among these stakeholders and the ones we are still not including in the discourse (enlarging the epistemology of our research) is vital to adapt and refine policy approaches continuously. This engagement is crucial for adjusting to technological and economic developments, and especially for aligning with the shifting economic, social, cultural, and natural landscapes across Europe (Bridge, 2018).

In conclusion, this book presents a compendium of current research, but also acts as a guide for an evolution of EU energy policy. It accentuates the necessity for collaborative interdisciplinary research in formulating policies that deeply consider the plurality of epistemological perspectives and cultural values. We hope that future energy policies will stretch this process and include the visions of non-human stakeholders in the energy transition, recognising that the discourse around technological and economic viability must move beyond mere trade-offs to embrace a more integrative approach. This shift is vital for ensuring a transition that is genuinely equitable and respects the diverse beliefs and stances that characterise individual and collective transformations.

REFERENCES

Arrobbio, O., & Sonetti, G. (2021). Cinderella lost? Barriers to the integration of energy social sciences and humanities outside academia. *Energy Research & Social Science, 73*, 101929. https://doi.org/10.1016/j.erss.2021.101929.

Bina, O., Fokdal, J., Chiles, P., Paadam, K., & Ojamäe, L. (Eds.). (2021). The inter- and transdisciplinary process: A framework. In J. Fokdal, O. Bina, P. Chiles, L. Ojamäe, & K. Paadam (Eds.), *Enabling the city* (pp. 17–33). New York: Routledge.

Bridge, G., Barca, S., Özkaynak, B., Turhan, E., & Wyeth, R. (Eds.). (2018). Towards a political ecology of EU energy policy. In C. Foulds, & R. Robison (Eds.), *Advancing energy policy: Lessons on the integration of social sciences and humanities* (pp. 163–175). Cham: Palgrave Macmillan.

Silvast, A., & Foulds, C. (2022). *Sociology of interdisciplinarity: The dynamics of energy research.* Cham: Palgrave Macmillan.

Sonetti, G., Arrobbio, O., Lombardi, P., Lami, I. M., & Monaci, S. (2020). "Only social scientists laughed"': Reflections on social sciences and humanities integration in European Energy Projects. *Energy Research & Social Science, 61*, 101342. https://doi.org/10.1016/j.erss.2019.101342.

Afterword 2: Considering the Role of the Scientific Community by Henry Jeffrey and Kristofer Grattan

Henry Jeffrey is a specialist in energy roadmaps and strategies for the offshore renewable energy sector. He is a co-director for the UK SuperGen (Offshore Renewable Energy) project and heads the Policy and Innovation Group in the University of Edinburgh's Institute for Energy Systems. Henry chairs the European Energy Research Alliance Ocean Energy Joint Programme and collaborates on numerous European ocean energy projects.

***Kristofer Grattan** is a Research Associate in offshore renewable energy at the University of Edinburgh, working within the Policy and Innovation Group, specialising in the production of policy guidance reports and energy system scenarios.*

From escalating geopolitical instabilities rising across the globe, to the increasingly shared lived experiences of the climate emergency, we exist now within an era of concurrent crises, where the time to act decisively is diminishing. With the provision of clean, sustainable, and just sources of energy for all of society residing at the intersection of these challenges, this book has arrived at a pivotal moment.

Across its chapters, it asks the reader to question: what is the role of the scientific community in fostering the ongoing transformation to sustainable and climate-resilient energy systems? From reading, it is clear that this question could be answered in a number of ways.

The energy transformation that is required will be an undertaking fraught with complexity, where the role of the scientific community in clearly interpreting, outlining, and communicating the challenges we can expect is of critical importance. Ensuring that all of society—from high-level policymakers to base-level consumers—can make informed and un-biased decisions is a necessary first step on this pathway to transformation. Its consequences are also clear, as evidenced through the focus on stakeholder engagement by Buylova et al. (Chapter 10), where the removal of knowledge barriers is identified as key for encouraging stakeholders to make positive energy efficiency decisions.

Equally, there is also a role for the scientific community in ensuring that the tools required to underpin this transition, such as the focus on energy system modelling by Süsser et al. (Chapter 11), are used in an open, accessible and inclusive manner. By embracing an interdisciplinary approach, which unites the wide range of experts that will be required to facilitate the energy transformation, it will be possible to deliver change that is empowering and just for those who need it most.

Finally, the role of technical experts in continuing to deliver the innovation, development, and deployment that is the natural conclusion of the energy transformation cannot be understated. However, by combining these roles with the insight and access that can only be achieved through comprehensive engagement with communities and local stakeholders, we can continue to strengthen support and engagement at a grassroots-level.

What is clear across these chapters, however, is that the slow rate of change associated with business-as-usual policymaking will not deliver the radical energy system transformation that is required. As we reach the midpoint for the critical decade for climate action, it has never been more important to embrace new approaches. From the use of machine learning and drones to monitor our expanding energy systems, to new configurations of energy governance, input and self-determination, there is an immediate and pressing need to break from the status-quo that has delivered us to this point. The ongoing energy transformation will be the greatest challenge of our generation, but it also presents the chance to radically re-imagine and re-design how energy interacts and shapes our everyday lives.

Our experience of working at the intersection of academia and energy policy within the ocean energy sector has highlighted to us the importance of access to strong networks of collaboration, supported by international organisations with the foresight and drive to set ambitious targets. However, we are also witnessing how powerful ideas, combined with considerate policymaking, can lift up entire communities and regions, often considered 'remote' from centres of high-tech industry. In many cases, these transformations are being driven from the 'bottom-up', where the societies whose landscapes are being re-shaped and re-defined, who come into daily contact with our new energy sources, are playing an active role.

It is this active role—i.e. the opportunity to become active participants in our own energy democracy—that is now most important of all. The consequences of allowing the climate emergency to continue unchecked have never been presented more clearly than they are today, yet the collective response of international governments is to move slowly and cautiously. To deliver the radical change required it is our shared responsibility as the scientific community, policymakers, and responsible consumers of energy, to encourage bold decision-making, accelerate our response and lend collaborative voices to securing a just and sustainable future for all.

Afterword 3: Empowering the Energy Transition: Collaborative Pathways Ahead for European Union Policy by Emma Bergeling

Emma Bergeling works as a Junior Policy Analyst in the Climate and Circular Economy team at the Institute for European Environmental Policy (IEEP). With a specialisation in Ecological Economics, her research interests span sustainable resource management, circular economy policy, and redesign of economic systems to fit within planetary boundaries.

The imperatives for rapidly mitigating climate change are difficult to exaggerate (EEA 2024). Indeed, the European Union (EU) has made significant strides with the European Green Deal, including the Fit for 55 policy package, and the REPowerEU plan (European Commission, 2019, 2021, 2022). However, sustained, focused efforts are required to align EU policy with the best available science and turn it into action on-the-ground. In this regard, this book offers important insights.

What is striking when reading the chapters is the remarkable range of topics covered. As an Environmental Policy Analyst, I was not initially expecting (although was delighted) to read, for example, about cyber resiliency in the context of the European energy transition, as discussed by Mersni et al. in Chapter 9. All chapter topics in turn contain a myriad of perspectives and considerations, underscoring the importance of collaborations between different spheres of knowledge to successfully navigate the complexity of the energy transition.

On the theme of action and success, numerous ready-to-implement solutions exist. For instance, Macrorie et al. (Chapter 6) explain that building retrofits can achieve at least a 60% energy reduction. Further potential for the construction sector lies in the newly revised Energy Performance of Buildings Directive specifying that all new buildings should be zero-emission as of 2030 (European Union, 2024). Similarly, Büscher et al. (Chapter 4) underscore the untapped potential of geothermal energy for community development. Such chapters offer vital steps to ensure that this potential is harnessed into energy policies with tangible environmental and social benefits.

A red thread running through the chapters is the notion of participation and empowerment for a just transition. This—the *how* of the energy transition—is central. For instance, Süsser et al. (Chapter 11) highlight gaps in energy modelling concerning three dimensions of justice: recognitional (whose worldviews are represented versus excluded?), procedural (who was involved in the process?), and distributional (how are gains and losses distributed?). The authors outline the important role that Social Sciences and Humanities (SSH) can play in addressing the gaps. Indeed, there is much room for behavioural, societal, and cultural interventions to play in the various stages of the EU policy process (Urios et al., 2022).

So, what else, in addition to the policy highlights in each chapter, can be derived from this book for EU policy? I would like to draw attention to the need for brave leadership. Buylova et al. (Chapter 10) highlight that some political groups use societal concerns regarding equity and cost of living as an argument for lowered ambitions in climate and environmental policies. This is a dangerous path to tread that fails to recognise the inherent interconnectedness of social, economic, and environmental issues. The cost-of-living crisis will certainly not be solved by runaway climate change. EU citizens know this, with a vast majority demanding increased political action on climate change (Andre et al., 2024). EU policymakers should realise this strong support and act accordingly. Specifically, setting science-based climate targets for 2040 based on the precautionary principle and equity will be one of many important tasks for the next legislators.

In a time of increasing polarisation and political opportunism, radical collaboration, and rethinking are more central than ever. Can you imagine a world where our economic systems are aligned with Earth's systems? A world where sustainability is no longer about mitigating harm, since the economy is designed to be as regenerative and distributive as nature itself? Table 4.1, by Büscher et al. (Chapter 4), gives a glimpse of what energy sources are present in such a future: wind, rain, agricultural waste, sunshine, Earth's deep heat. Not a sign of harmful fossil fuels.

No-one said mitigating climate change would be easy; it demands change across all of society. But science is shouting loud and clear that we have everything to win by embarking on this journey with speed and determination. So, let us proceed to work in our different roles; in Science, Technology, Engineering and Mathematics (STEM) and SSH, in our communities and in EU policy, apart and—crucially—together. All with the common aim of thriving alongside other species within the limits of the green and blue planet we have the great privilege of calling our home.

Acknowledgements

Many thanks to my much-cherished colleague Chiara Antonelli for her invaluable exchange of ideas, support, and continuous feedback. Work on this Afterword was funded by IEEP's involvement in the SHARED GREEN DEAL project, which is funded under the EU Horizon 2020 research and innovation programme (grant no. 101036640).

References

Andre, P., Boneva, T., Chopra, F., & Falk, A. (2024). Globally representative evidence on the actual and perceived support for climate action. *Nature Climate Change, 14*(3), 253–259. https://doi.org/10.1038/s41558-024-01925-3.

European Commission. (2019). *The European Green Deal.* COM(2019) 640 final. Brussels: European Commission.

European Commission. (2021). *'Fit for 55': delivering the EU's 2030 Climate Target on the way to climate neutrality.* COM(2021) 550 final. Brussels: European Commission.

European Commission. (2022). *REPowerEU Plan.* COM(2022) 230 final. Brussels: European Commission.

European Union. (2024). *Directive (EU) 2024/1275 of the European Parliament and of the Council of 24 April 2024 on the energy performance of buildings (recast).* Brussels: Official Journal of the European Union.

EEA [European Environment Agency] (2024). *European climate risk assessment—Executive summary.* EEA Report No 1/2024. Luxembourg: Publications Office of the European Union. https://doi.org/10.2800/204249.

Urios, J., Casert, C., Gore, T., Foulds, C., Afghani, N., (2022). *Behavioural, cultural and social issues in EU Green Deal policy documents*. Cambridge: SHARED GREEN DEAL.

Afterword 4: Reinserting the Missing Piece: Integrating the Human Dimension in Energy Policy by Rod Janssen and Audrey Nugent

Rod Janssen *is an expert in sustainable energy policies with a special focus on energy efficiency. He is President of the Brussels-based not-for-profit organisation, Energy Efficiency in Industrial Processes; and Editor of the blog on sustainability energy issues, Energy in Demand. Most of his career has been in Europe, helping governments, associations, and civil society pursue their path to a sustainable energy future.*

Audrey Nugent *is the Global Advocacy & Campaigns Director at the World Green Building Council (WorldGBC). In this role, Audrey leads on global advocacy and campaigns collaborating with Green Building Councils around the world to champion ambitious and innovative public policies. Audrey has over 15 years of experience working in policy and advocacy and has worked across private, public and NGO sectors.*

As a Swedish colleague once said: while energy efficiency is not difficult, it *is* complicated. The same, we could say, goes for the whole energy transition. Unfortunately, many circles (still) default to the position that technology will solve all this complexity. The wide-ranging discussions in this book show that that default position does not work and it is time to adopt a systemic approach to the climate crisis we face.

Already in 1984, the US National Research Council wrote *Energy Use: The Human Dimension*. This book discusses the major social and political components of energy policy to put energy into its human context. This has influenced our thinking ever since, but its impact was much less than it should have been. Now, 40 years later, as the climate crisis accelerates, we are finding an entire new generation that is understanding the need to embed the human. All of the interdisciplinary teams contributing to this book address the failure to do so until now. Each chapter exposes a shortcoming in our energy policymaking and implementation. In doing so—and in proposing remedies—together they, thus, move us into the right direction. They make you *think differently* about the energy transition than the default approach. They bring the human dimension front and centre, showing the complexity of energy-related decisions.

These chapters clearly demonstrate that bringing in the human dimension requires an interdisciplinary approach. Social Scientists and Engineers need to collaborate to be able to develop policies and programmes that lead to optimal outcomes for a liveable climate.

A prevalent theme is the importance of community-led participation—whether that be related to the implementation of EU policy measures (Mey et al., Chapter 2; and Macrorie et al., Chapter 6), engaging society for better acceptance of less established (or understood) energy sources (Büscher et al. Chapter 4), or navigating digital infrastructures (Buylova et al., Chapter 10). Specifically, such contributions reiterate how a human-centric approach enables collaboration with local communities to build common ground and support implementation.

The success of the energy transition is contingent on community-led implementation and engagement that puts consumers at the heart of the transition—a bottom-up policy-making process, if you will. This is particularly important for the built environment, which has often been overlooked due to the (mis)perceptions of how challenging it would be to decarbonise. Now, it is true that it is challenging—and that is because of the human factor. Engineering solutions keep tripping over it, for having missed it in their plans. It takes technical solutions to improve the energy performance of the building envelope, but it takes the human factor to have that work actually undertaken.

Europe is falling far short in undertaking the energy renovations needed to meet the objectives of the EU Renovation Wave Strategy (European Commission, 2020). One key element in accelerating the pace is providing appropriate information for consumers to make good decisions. As Calver et al. (Chapter 5) states, for citizens to participate in domestic energy actions, there is the need to relate alternative practices to their individual situations, and see these practices in relation to both co-benefits and drawbacks, such as finance and health outcomes.

One-stop-shops are a great instrument in this regard (Macrorie et al., Chapter 6), as they can create a consumer journey for improved energy performance that balances needs and capabilities of the consumer.

Our clean energy transition is a challenge never experienced before. The chapters on geothermal, nuclear, societal engagement, social acceptance, communities, and the buildings directive all show that we will not meet this challenge if we do not take the consumer or citizen as the starting point. The interdisciplinary approach that follows from that starting point should be the new default position.

REFERENCES

European Commission. (2020). *A renovation wave for Europe—Greening our buildings, creating jobs, improving lives.* COM(2020) 662 final. Brussels: European Commission.

US National Research Council. (1984). *Energy Use: The Human Dimension.* Washington, DC: The National Academies Press. https://doi.org/10.17226/9259.

Index

A
Acceptance, x, 36
Advanced Persistent Threats (APTs), 119, 120
Africa-EU Energy Partnership (AEEP), 55
African Union (AU), 53, 170
Agriculture, 91, 93–96
Agrivoltaics (AV), 91–93, 103–110
  deployment, 92, 94–97
Agrivoltaic technology, 9
Apprenticeships, 77, 79, 80
Arrobbio, Osman, 170
Artificial Intelligence (AI), 117, 120, 121, 124–126
Assessment, 21–23, 26
Assumptions, 148–153
Austria, 23, 79

B
Bavarian Energy Atlas, 25
Bavarian Ministry of Economic Affairs, Regional Development and Energy, 25
Behavioural Science, 150, 151

Bergeling, Emma, 177
Bottom-up, 36, 38, 39, 175, 182
Building Energy Management System (BEMS), 135
Building Renovation Plans, 75, 77
Business-as-usual policymaking, 174

C
Capabilities, 77
Citizen Energy Community (CEC), 19, 20, 26
Citizen engagement, 33
City Building, 77
Civic Square, 77
Civil Engineering, 5
Clean energy for all Europeans, 21
Climate scepticism, xi
Co-creation, 40
Co-design, 150
Co-learning, 47, 50, 53
Colouring Cities Research Programme (CCRP), 21
Common Agricultural Policy (CAP) framework, 95
Commons, 54

Communication expertise, 152
Community energy, 8, 19, 21–23, 26
Community engagement, 95, 152
Community participation, 7, 8
Computer Engineering, 147
Computer Science, 104
Conflict, 24
Construction, 75, 76, 78–81
  management, 76
Construction Project Management, 63
Contextual factors, 106
Controversy, 104
Cooling, 35, 37, 39, 41
Cost, 133–138, 140
Costs and benefits, 35
Cross-fertilisation, 7
Crowdfunding, 39
Cyberattacks, 117–120, 122
Cyber espionage, 119
Cyber resilience, 123
Cybersecurity, 117, 118, 120–126
Cyberthreats, 117–120, 122, 123, 125
Czechia, 79
  Ministry of Industry and Trade, 79

D

Data protection laws, 125
Decarbonisation, 4, 133, 137
Decentralised, 48, 55
Decision-making, 134, 135, 139
Decision-making processes, 8
Deliberation, 36, 38, 40
Delivering the European Green Deal package, 4
Democratic, 19, 25
Denmark, 25, 36
Depoliticise, 149
Developer, 35, 36, 38
Development, 47–49, 51–54
  cooperation, 49

Dialogue, 8, 10
Digital energy infrastructure, 9, 118, 125
Digital energy systems, 125
Digitalisation, 117
Digital planning tools, 26
Direct Labour Organisations (DLOs), 77, 80
Disciplines, 166, 167
Diverse, 76, 80, 81
Document analysis, 118
Donor, 49, 53, 54

E

East African Rift System (EARS), 49, 51
Eastern Africa, 50, 52
Economics, ix, xiii, 20, 21, 24, 25, 91, 94, 96
Economic vulnerabilities, 150
Electrical and Electronics Engineering, 118
Employment, 76, 78, 80, 81, 152
Enabling environment, 54
Energy, 33–40, 75, 77, 79, 81
  efficiency, 75, 79
  future of energy, 37, 38
  futures, 151
  justice, 148, 149
  system models, 147, 150, 152
  System Optimisation Models (ESOMs), 147
Energy Communities Repository, 22, 23
Energy community, 53–55
Energy democracy, 175
Energy efficiency, 133–141, 169, 174, 181
Energy Efficiency Directive, 5
Energy justice, 9, 52
Energy literacy, 8, 62, 64, 65, 68–70

Energy production, 103, 104, 108, 109
Energy-SHIFTS, 7
Energy transition, 102, 110
Energy Union, 4
Engagement, 54
　inclusive engagement, 38, 40
　societal engagement, 33–40
　top-down approaches to engagement, 34
Engineering, 50, 63
　Civil Engineering, 76
　Structural Engineering, 76
Environmental sociology, 33
EPBD
Ethical AI practices, 126
EU-Africa Green Energy Initiative, 55
European Climate Law, 4
European Commission (EC), v, 47, 48, 54, 55
　Climate-Neutral and Smart Cities Mission, v
　Directorate-General for Energy, 54
　Directorate-General for International Partnerships, 53
　Directorate-General for Research and Innovation, v
European Construction Blueprint, 80
European Cyber Crises Liaison Organisation Network, 124
European Federation of Citizen Energy Cooperatives, 19
European Green Deal, xi, 33, 40, 147
European Investment Bank (EIB), 54
European Union (EU), 19–23, 25, 26, 47–51, 53–55, 62–64, 66–70, 91, 93, 95, 96, 147
　Artificial Intelligence (AI) Act, 120, 125
　building stock, 75
　Cities Mission, 133
　Clean Energy for all Europeans, 4
　cross-sector cooperation energy operators, 123
　cybersecurity legislation, 123, 124
　digital energy infrastructure, 118, 125
　Energy Efficiency Directive, 5, 79, 133
　Energy Performance of Buildings Directive (EPBD), 64, 75, 133, 135, 140, 178
　Energy Union, 4
　Fit for 55, 153, 177
　Green Deal, 91, 177
　Just Transition Mechanism, 153
　Network and Information Security 2 Directive (NIS2 Directive), 121, 124
　Network and Information Security (NIS Directive), 121
　Network Code, 121, 124
　performance of buildings directive, 5
　Renewable Energy Directive, 5, 22
　Renovation Wave, 69
　Renovation Wave Strategy, 4, 182
　Skills Registry, 79
　Solar Energy Strategy, 22
Everyday life, x
Expert elicitation, 33, 34
Exploration tools, 153

F
Fairness, 147, 150, 153
Financial
　obstacles, 47
　resources, 54
　support, 54
Financial challenges, 35
Fit for 55, 5, 153, 177
Framing, 65, 67
France, 39

Funding, 170
Future, vii, xi, xiii, 10, 36, 37, 63, 70, 104, 111, 118, 120, 125, 147, 150, 152, 153, 166, 171, 175, 179

G
GDP growth, 147
Geodata, 25
Geographic Information Systems (GIS), 21, 24, 25
Geographic Science, 20
Geophysics, 33
Geoscience, 33, 34, 50
Geothermal, 33–41
Geothermal energy, 8
  community, 51–54, 170
  development license, 52, 54
  potential, 47–49, 51, 53
German Renewable Energy Act, 24
Germany, 25, 39
Governance, 75, 104, 106, 107, 109
Grattan, Kristofer, 173
Greater Manchester, 63, 67, 68
Green Deal, 177
Greenhouse gases (GHG), 133, 136–138, 140

H
Heating, 35, 37, 39, 41
Her Own Space, 78
Homes, 66, 67
Horizon2020, 6
Housing, 77, 81
Housing association, 133, 134, 136–138, 140, 141
Human error, 120, 126
Human geography, 33, 63, 76, 148
Hungary, 105, 106

I
Iceland, 38
Inclusive, 76, 78, 80, 81
  pathways, 76, 81
Inclusiveness, vi
Inclusivity, 76, 77, 170
Indexification of poverty, 151
Industry-community relationships, 38
Information, 63–70
Innovation, 104, 109
Institutional allies, 23, 26
Integrated Assessment Models (IAMs), 147
Integrated policy design, 96
Interdisciplinary, vii, x, xi, xiii, xiv
Intermediaries, 23, 24, 26
International Renewable Energy Agency (IRENA), 147
International Thermonuclear Experimental Reactor (ITER), 107
Internet of Things (IoT), 117
Italy, 36

J
Janssen, Rod, 181
Jeffrey, Henry, 173
Justice
  distributional, 148, 152
  procedural, 148, 149, 151
  recognitional, 148, 150, 151
Just pathways, 153
Just transition, 18, 19, 26, 146–148, 152, 153, 178
Just Transition Mechanism, 153
JustWind4All, 147

K
Kenya, 47
Knowledge(s), 6–9, 62–67, 69, 70

INDEX 189

L
Landlords, 77
Law, 5, 20, 21, 107, 108, 124
Leadership, 178
LEAP-RE program, 47
Least-cost pathways, 147, 150
Legal framework, 19, 23
Liberalisation, 52
Literature review, 92, 95
Local, 77
  area, 75
  partnerships, 77
Local authorities, 37
Local Energy Advice Demonstrator (LEAD), 67
Local government, 23, 24, 26
Low-carbon, v–vii, x, xiii, 4, 6, 8, 10, 35, 63, 64, 66, 67, 69, 91, 103–106, 108–111
Low-carbon energy projects, 103, 105, 109

M
Maintenance, 77
Marginalised voices, 149
Minority groups, 75
Modelling, 140, 147–153
Multicriteria decision-making models, 169
Multicriteria model(s), 9, 134, 135, 138, 140, 141
Multi-layered security approach, 125
Municipality, 22, 76–78

N
National Building Renovation Plans, 133
National Energy and Climate Plans (NECPs), 23, 64
National target, 22
Network Code, 121

NEWCOMERS project, 64
New European Bauhaus, 140
Newspaper articles, 104, 108
Nuclear fission, 107, 109
Nuclear fusion, 103–110
Nugent, Audrey, 181

O
Observatories, 110
Ocean energy, 174
Off-grid, 47, 55
Offshore wind, 102–110
One-stop shops (OSS), 23, 76, 78, 183
Operator, 34, 36, 38–40
Opportunities, 78, 80, 81

P
Pareto frontier, 136–139, 141
Participation, 62–69
Participatory, 149, 150, 152
  dialogue, 153
  modelling, 153
Participatory approach, 50, 54
People Powered Retrofit, 80
Perceptions, 37, 52
  societal perceptions, 35
Phishing attacks, 119
Physics, 104
Place-based, 76–78, 80, 81
Planning, 78
Planning tools, 24, 25
Policy research, 148
Political Science, 20, 134
Power relations, 105
Predictive Modelling, 33
Predictive threat identification, 126
Privatisation, 52
Procurement, 77
Property regime, 54
PROSEU, 21

Prosumership, 19
Psychology, 20, 148, 151
Public, 36, 37, 40
  awareness, 36, 41
  deliberation exercises, 36, 37
Public perception, 106
Public Policy Studies, 118

Q
Qualitative research, 92
Questionnaire, 92

R
RED II, 22, 26
Regulation, 75, 76, 79
Renewable energy, 102, 103, 107
Renewable Energy Community (REC), 19, 20, 26
Renewable Energy System technology, 91
Renewable energy technologies, 33, 40
Renocally, 77
Renovation, 4, 66, 68, 75, 78, 133–135, 137, 138, 140, 182
Repair, 77
REPowerEU, 5, 22, 33, 40, 62, 91, 102, 116, 177
REScoop, 19, 64
Research and Development (R&D), 96
Research and innovation (R&I), 6, 7
Research-policy interface, 165
Resilience, 124
Retrofit, 9, 75–81, 170, 178
Revised Energy Directive, 75, 133, 135, 140, 178
Risks, 33–35, 37, 38, 41
Rural Energy Communities Advisory Hub, 22

S
Scale, 165
Science Communication, 33
Science, Technology, Engineering and Mathematics (STEM), vi, vii, xiii, xiv
Scientific community, 173–175
Scotland, 77
Security Studies, 118
SEEDS, 147, 151
Seismicity, 35
Semi-structured interviews, 92
Sentiment analysis, 104, 108, 109
SENTINEL, 147, 151
SHAPE ENERGY, 7
SHARED GREEN DEAL, 179
Skilled retrofit workforce, 76
Skills, 23, 75–81, 170
Smart grids, 117, 120
Smart grids and Supervisory Control and Data Acquisition (SCADA) systems, 119
Social acceptability, 9, 102–106, 108–111, 170
Social Climate Fund, 70
Social dynamics, 50
Social Geothermal Sciences, 33
Social License to Operate, 38, 40
Social-political storylines, 151
Social Sciences, 50
Social Sciences and Humanities (SSH), v–vii, x, xi, xiii, xiv
Sociology, 76, 96
Sociology of Interdisciplinarity, 170
Socio-technical, 76, 81, 91
Solar PV, 97
Sonetti, Giulia, 170, 171
Spain, 39
Spillover, 37
SSH CENTRE, vii, xiv, 170
SSH fragmentation, 7
Stakeholder cooperation, 123

Stakeholder participation, 135, 140
Stakeholders, 103, 105, 107, 108, 110, 111
Strategies, 80
Structural Engineering, 76
Subsurface, 48, 50–52, 54
Subsurface exploration, 33
Supply chain, 75–78, 80, 81
Sustainability challenges, 4
Sweden, 135, 136, 138
Switzerland, 37

T

Technical assessments, 37
Technical assistance, 110
Technical experts, 174
Techno-economic, 147–149
Techno-managerial, 151
Telecommunications Systems and Networks, 118
The Green Register, 80
Topic modelling, 104, 108
Trade-offs, 133–135, 137, 139, 140
Training, 75–81
Training programmes, 9, 110, 118, 122, 126
Transdisciplinary debate, 153
Transect walk, 51

Transition impacts, 149, 150
Transparent, 148–151, 153
Trust, 35, 36, 38, 39, 65, 68–70

U

UK, 35, 39
UK Research and Innovation, xv
Uncertainties, 149, 151, 152
Unionisation, 78
Union of Construction, Allied Trades Technicians (UCATT), 78
United Kingdom (UK), 75–78, 80
US National Research Council, 181

V

Van der Vlies, Rosalinde, v
Visioning documents, 151
Visualisation, 152
Vocational education and training (VET), 75

W

Wellbeing, 134, 139
Women Can Build project, 80
Workshop, 47, 50–52, 63, 65, 67, 68, 134, 135, 138, 140
Worldviews, 148–150

Printed by Printforce, the Netherlands